FREE BOOKS

www.*forgottenbooks*.org

You can read literally thousands of books for free at www.forgottenbooks.org

(please support us by visiting our web site)

Forgotten Books takes the uppermost care to preserve the entire content of the original book. However, this book has been generated from a scan of the original, and as such we cannot guarantee that it is free from errors or contains the full content of the original. But we try our best!

Truth may seem, but cannot be:
Beauty brag, but 'tis not she;
Truth and beauty buried be.

To this urn let those repair
That are either true or fair;
For these dead birds sigh a prayer.

Bacon

The Locomotive Publishing Co., Ltd.

3 AMEN CORNER .·. LONDON, E.C.4

RAILWAY SIGNAL AND PERMANENT WAY ENGINEERS' POCKETBOOK. 260 pp. Price 5s. Postage 3d.

GENERAL CONTENTS—

Railway Signalling.—Development of Fixed Signals—Board of Trade Requirements in regard to Signalling on Railways—Description of Various Types of Signals and Purposes for which used—Signal Boxes—Signal Lighting—Signalling Systems—Track Circuits and Automatic Signals—Automatic Block Signals—The Block System, Cab Signalling and Automatic Train Control—Telephones on Railways—Telegraphy.

Permanent Way.—Board of Trade Regulations—Rails—Sleepers —Chairs — Keys — Fishplates—Ballasting — Super-Elevation and Tables—Gauging and Checking of Curves—Switches and Crossings —Setting out of Permanent Way—Calculation of Switch and Crossing Angles—Improvement of Existing Curves—Electrical Equipment of Railway Tracks relative to Permanent Way Maintenance. Useful Tables, and Metrical Equivalents.

An Indispensable Compendium for the Engineer, Draughtsman, Inspector, Lineman Ganger, etc., etc.

SUPERHEATING ON LOCOMOTIVES. By J. F. GAIRNS. Second Edition, 120 pp. (Illustrated.) Price 3s. 6d., Postage 3d.

The Why and the Wherefore of Superheated Steam. Advantages —Economics and Special Problems—History and Development of the Superheater Locomotive—Principal Superheaters in Use— The Maintenance of Superheater Locomotives.

LOCOMOTIVE RUNNING SHED NOTES. 98 pp. (Illustrated.) Price 3s. 6d., Postage 3d.

Locomotive Running Shed Management—Training of Enginemen —Locomotive Defects—The Cure of Bad Steaming Engines—The Slipping of Driving Wheels—Running Shed Breakdown Tools and Tackle—Testing Locomotive Valves and Pistons in Steam—How to Treat Big Ends—Hydraulic Wheel Drops for Running Sheds— Re-Metalling and Fitting Eccentric Strap Liners—Regrinding Steam Cocks—Refitting Big End Brasses and Axle Brasses— Protecting Boiler Plates—Brake Rigging, etc.

NOTES ON RAILWAY SIGNALLING. By J. PARSONS, A.M.I.S.E., and B. W. COOKE, A.M.I.S.E. 74 pp. (Illustrated.) Price 3s. 6d., Postage 3d.

Written with the object of giving in a concise form a general idea of the methods and appliances adopted in Great Britain—Evolution of Signalling—Repeating and Indicating—Shunting Signals— Connections to Points—Compensation—Facing Point Locks— Locking Apparatus—Signal Boxes—Level Crossing Gates—The Block System—Single Line Working—Track Circuit—Fog—Cab Signals, etc.

SEND FOR COMPLETE LIST OF RAILWAY PUBLICATIONS POST FREE.

PITMAN'S TECHNICAL PRIMER SERIES

Edited by R. E. NEALE, B.Sc., Hons. (Lond.)
A.C.G.I., A.M.I.E.E.

STEAM
LOCOMOTIVE CONSTRUCTION
AND MAINTENANCE

PITMAN'S
TECHNICAL PRIMERS

Edited by R. E. NEALE, B.Sc. (Hons.), A.C.G.I., A.M.I.E.E.

IN each book of the series the fundamental principles of some sub-division of engineering technology are treated in a practical manner, providing the student with a handy survey of the particular branch of technology with which he is concerned. They should prove invaluable to the busy practical man who has not the time for more elaborate treatises. Each 2s. 6d. net.

THE STEAM LOCOMOTIVE. By E. L. AHRONS, M.I.Mech.E.

BELTS FOR POWER TRANSMISSION. By W. G. DUNKLEY, B.Sc. (Hons.).

WATER-POWER ENGINEERING. By F. F. FERGUSSON, A.M.I.C.E.

PHOTOGRAPHIC TECHNIQUE. By L. J. HIBBERT, F.R.P.S.

HYDRO-ELECTRIC DEVELOPMENTS. By J. W. MEARES, F.R.A.S., M.Inst.C.E., M.I.E.E.

THE ELECTRIFICATION OF RAILWAYS. By H. F. TREWMAN, M.A.

CONTINUOUS CURRENT ARMATURE WINDING. By F. M. DENTON, A.C.G.I.

MUNICIPAL ENGINEERING. By H. PERCY BOULNOIS, M.Inst.C.E., F.R.San.Inst., F.Inst.S.T.

FOUNDRY WORK. By BEN SHAW and JAMES EDGAR.

PATTERN-MAKING. By BEN SHAW and JAMES EDGAR.

THE ELECTRIC FURNACE. By FRANK J. MOFFETT, B.A., M.I.E.E., M.Cons.E.

SMALL SINGLE-PHASE TRANSFORMERS. By EDGAR T. PAINTON, B.Sc., A.M.I.E.E.

PNEUMATIC CONVEYING. By E. G. PHILLIPS, M.I.E.E., A.M.I.Mech.E.

BOILER INSPECTION AND MAINTENANCE. By R. CLAYTON.

ELECTRICITY IN STEEL WORKS. By W. MACFARLANE, B.Sc.

MODERN CENTRAL STATIONS. By C. W. MARSHALL, B.Sc.

STEAM LOCOMOTIVE CONSTRUCTION AND MAINTENANCE. By E. L. AHRONS, M.I.Mech.E., M.I.Loco.E.

HIGH TENSION SWITCH GEAR. By H. E. POOLE, B.Sc., A.C.G.I., A.M.I.E.E.

HIGH TENSION SWITCH BOARDS. By the Same Author.

POWER FACTOR CORRECTION. By A. E. CLAYTON, B.Sc., A.K.C., A.M.I.E.E.

TOOL AND MACHINE SETTING. By P. GATES.

TIDAL POWER. By A. STRUBEN, O.B.E., A.M.I.C.E.

SEWERS AND SEWERAGE. By H. GILBERT WHYATT, M.I.Inst.C.E., M.Roy.San.Inst., M.Inst.M. and C.E.

ELEMENTS OF ILLUMINATING ENGINEERING. By A. P. TROTTER M.I.C.E.

COAL-CUTTING MACHINERY. By G. E. F. EAGAR, M.Inst.Min.E.

GRINDING MACHINES AND THEIR USE. By T. R. SHAW, M.I.Mech.E.

ELECTRO-DEPOSITION OF COPPER. By CLAUDE W. DENNY, A.M.I.E.E.

LONDON: SIR ISAAC PITMAN & SONS, LTD.

STEAM LOCOMOTIVE CONSTRUCTION AND MAINTENANCE

DESCRIBING WORKSHOP EQUIPMENT AND PRACTICE IN
THE CONSTRUCTION OF MODERN STEAM RAILWAY
LOCOMOTIVES, WITH NOTES ON INSPECTION,
TESTING, MAINTENANCE AND REPAIRS

BY

E. L. AHRONS

M.I.Mech.E., M.I.Loco.E.

LONDON
SIR ISAAC PITMAN & SONS, LTD.
PARKER STREET, KINGSWAY, W.C.2
BATH, MELBOURNE, TORONTO, NEW YORK
1921

PRINTED BY
SIR ISAAC PITMAN & SONS, LTD.
BATH, ENGLAND

PREFACE

THIS book describes in an elementary manner some of the processes which the principal parts of a locomotive undergo during construction, and may be considered as a companion volume to the primer on *The Steam Railway Locomotive* in this series.

The subjects of foundry practice and general machine shop processes being dealt with in other primers, they are here considered briefly and only in their special relationship to locomotive construction. Naturally, no attempt can be made to deal exhaustively with machinery and processes in so small a book, and those readers who wish for a more detailed treatment of the subject may be referred to *The Construction of the Modern Locomotive*, by George Hughes, M.Inst.C.E., M.I.Mech.E., which, though not of very recent date, remains the only standard work on British locomotive workshop practice, and contains a mass of valuable information.

At the request of the publishers a chapter has been added on the maintenance of the locomotive in service, which gives, also in an elementary form, a short account of the wear and tear of some of the more important parts, and some idea of the repairs required.

PREFACE

The thanks of the author are due to Mr. G. J. Churchward, M.Inst.C.E., Chief Mechanical Engineer of the Great Western Railway, and to the various manufacturers of special machine tools for the photographic illustrations. For a few other illustrations, including the method of erecting the locomotive cylinders and frames, the author is indebted to the kindness of The Locomotive Publishing Co., Ltd.

E. L. AHRONS.

NOTTINGHAM,
December, 1920.

CONTENTS

CHAP.		PAGE
	PREFACE	V
I.	GENERAL CONSIDERATIONS AFFECTING METHODS OF CONSTRUCTION	1
II.	BOILER SHOP	8
III.	FOUNDRIES	27
IV.	FORGINGS, SPRINGS, ETC.	35
V.	MACHINE SHOPS—FRAMES AND CYLINDERS	46
VI.	MACHINE SHOPS—AXLES AND WHEELS	56
VII.	GENERAL MACHINE SHOP	70
VIII.	ERECTING SHOP	80
IX.	SETTING THE VALVES	96
X.	INSPECTION AND TESTING	107
XI.	LOCOMOTIVE MAINTENANCE AND REPAIRS	116
	INDEX	133

ILLUSTRATIONS

FIG.		PAGE
1.	Diagram illustrating the manufacture of locomotive engine and tender . .	2, 3
2.	Sections through locomotive boiler .	10
3.	Radial drilling machine for boiler plates .	12
4.	Diagram of plate-bending rolls . .	13
5.	Boiler shell drilling machine . . .	14
6.	Hydraulic flanging press . . .	16
7.	Completed locomotive boiler . .	25
8.	Axlebox moulding machines and moulds, G.W.R. Iron Foundry, Swindon .	30
9.	Iron foundry, Swindon Locomotive works	33
10.	Blocks for forging straight axles . .	36
11.	Stages in forging a crank axle . .	38
12.	Buffer forging	40
13.	Forged buffer casing . . .	42
14.	Pieces for forging buffer casing . .	42
15.	Stamping shop with drop hammers . .	43
16.	Underhung laminated spring . .	44
17.	Frame plate	46
18.	Punching the outline of the frame plate .	47
19.	Marking-off cylinders . . .	49
20.	Marking-off cylinder port faces . .	50
21.	Double cylinder-boring machine . .	51
22.	Horizontal drilling and facing machine for cylinders	53
23.	Straight axle	56
24.	Centre driven axle lathe . . .	57
25.	Crank sweep milling machine . .	59
26.	Horizontal turning and boring mill for wheel centres and tyres . .	61

ix

ILLUSTRATIONS

FIG.		PAGE
27.	Boring tyres on horizontal boring mill .	62
28. 29.	Tyre fastenings	63
30.	Locomotive wheel lathe . . .	65
31 to 34.	Tyre gauges	66
35.	Wheel balancing machine, Great Western Railway, Swindon	68
36.	Cast iron cylinder for piston rings. .	72
37.	Tools in turret for machining piston rings	72
38.	Grinding a piston rod . . .	74
39.	Crosshead and slide blocks . . .	75
40.	Milling connecting rods . . .	77
41.	Trammelling the frames . . .	81
42.	Erecting locomotive frames and cylinders	83
43.	Marking the valve spindle for valve setting	100
44.	Finding the dead centres . . .	102
45.	Locomotive testing plant, Great Western Railway, Swindon . . .	114
46.	Oval tube holes and cracked tube plate.	119
47.	Patch on firebox plate . . .	121
48.	Bushed tube holes in tube plate . .	122
49.	Cracked and broken stays, with heads wasted away inside firebox. . .	123
50.	Tread of tyre showing wear and section after re-turning	127

STEAM LOCOMOTIVE CONSTRUCTION AND MAINTENANCE

CHAPTER I

GENERAL CONSIDERATIONS AFFECTING METHODS OF CONSTRUCTION

A LARGE volume would be required to deal fully with the manufacture of the many different parts of a locomotive and to describe the various processes in the foundries, smithy and forge, boiler-shop, machine shop, and erecting shops. Moreover the methods of construction of the various parts differ considerably in a small works, where perhaps only one or two engines are built at a time, from those in a large works, where from a dozen to fifty engines of the same class may be built to a single order. In the first case the methods and machines of an ordinary well equipped engineering shop would be used to a large extent, but in the second case there are employed special templates, " jigs," fixtures, and above all, special machinery adapted for certain definite operations. The cost of such special machinery and appliances requires a large

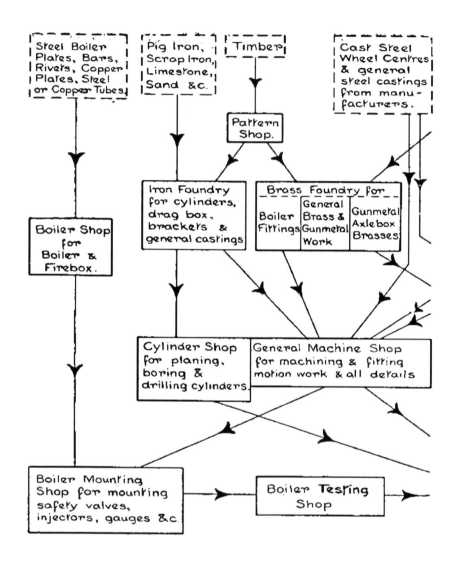

Fig. 1.—Diagram Illustrating the Manufacture of Locomotive Engine and Tender.

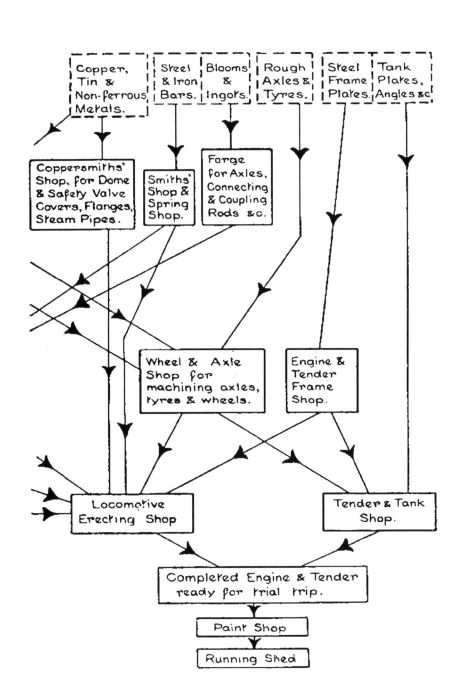

4 STEAM LOCOMOTIVE CONSTRUCTION

capital outlay, which could not profitably be expended by a small works. It pays to make the necessary special time and labour-saving tools and appliances only when several similar engines are built together, or, as in many railway works, when much of the "motion work" and other parts of one class of engine are interchangeable with those of another class. Naturally when the conditions are such that a large number of similar engines can be constructed together with the help of special tools and "jigs," the cost per locomotive is very considerably reduced. An isolated engine of a new design is necessarily expensive, but it generally pays a railway company to build such an engine, so that it can be tested thoroughly, and the necessary modifications and alterations made before placing a large order.

Progress of Work through the Workshops. A general idea of the order in which work passes through the various shops may be gained from Fig. 1, beginning with the raw materials. The latter term may and generally does include a considerable quantity of semi-finished material. Steel castings, axles, and tyres may be mentioned specially in this connection. One or two locomotive works, both of the railway companies and private firms, make their own steel castings, but generally these are purchased from outside manufacturers. The same applies to rough forged axles. Tyres are, in most cases, rolled at the

GENERAL CONSIDERATIONS OF CONSTRUCTION 5

mills of the steel makers who make a speciality of such work.

The diagram Fig. 1 explains itself, but there are considerable variations in the practice at different works. For instance, boiler mounting work is frequently done in an annexe of the boiler shop itself. Of general machine shops there are frequently two or even three, one or two being devoted to the large machines employed upon the heavier parts, and another to the smaller machines for small pieces and repetition work. The diagram does not show the route taken by every part which goes to form the locomotive. To do this would be impossible. For instance, the firebox of a locomotive boiler is stayed with a large number of screwed steel and copper stays. In the latest practice these are screwed in machines which are placed in a convenient bay in the boiler shop itself, but it is more often the case that they are made in one of the machine shops, from which they go to the boiler shop. There are frequently many such cross paths for various details, and to show these would make the diagram unnecessarily complicated. Certain materials coming from the manufacturers have also been omitted, such as the asbestos mattresses or magnesia blocks used for boiler covering, or *lagging* as it is termed. These are taken from the stores direct to the erecting shop, where the boiler is lagged during the later stages of the erection of the engines.

One department, not shown on the diagram,

6   STEAM LOCOMOTIVE CONSTRUCTION

must be mentioned briefly. This is the template shop, in which thin sheets of metal are cut out to the finished shapes of the different parts of the boiler and engine details in accordance with the drawings. These templates are sent to the forge, boiler and machine shops, where they are laid upon the corresponding parts being manufactured, instead of marking off the latter to measurements by rule. Much time is saved by the use of the templates.

Stores and Costing Accounts · All raw materials, or semi-manufactured parts such as steel castings, are received from the makers into the general stores. The works manager issues the necessary orders to the stores for such materials or articles, which are sent to the various shops in which they are required. The cost of these is debited to the particular order, say for ten locomotives, for which they are to be used, each order receiving an order number, against which everything is charged that is required for that particular lot of engines, until they are completed. The cost of labour is determined by means of daily time sheets, on which the number of hours taken by each workman is entered against the order on which he has been engaged. Finally an overall charge is added for general expenses which include rent, rates, taxes, management, unskilled labour, drawing office expenses, power, lighting and depreciation of plant. This charge usually takes the form of a fixed percentage either of the

GENERAL CONSIDERATIONS OF CONSTRUCTION 7

total cost of materials and skilled labour combined, or of the cost of skilled labour alone, generally the latter. The combined total gives the cost of the engines built to the particular order under consideration, which divided by the number of locomotives, gives the cost per engine.

Before the war the cost of an average main-line engine and tender, when say ten were constructed together, would vary roughly from about £2,500 to £3,700 at the works, according to the size of the engine, and the fluctuating prices of materials. Post-war conditions have caused these costs to be from $3\frac{1}{2}$ to 4 times the above amounts.

CHAPTER II

BOILER SHOP

Boiler Plates and Making the Boiler Barrel.
The steel plates are delivered into the boiler shop
after they have been sheared to sizes about
$\frac{1}{2}$ in. larger all round than the finished sizes to
which they are put together in the boiler. They
must be of the very finest quality, since they have
to undergo processes, such as flanging, punching,
etc., which would inevitably crack or otherwise
injure inferior material.

After their arrival, the plates are first straightened to remove any slight waviness, by passing
them through a machine, sometimes known as a
" mangle," which has a number of hard steel
rollers, usually four above and three below. By
passing the plate through between the upper and
lower rollers two or three times the irregularities
are removed.

The length of the barrel of the boiler may vary
from about 10 ft. to $16\frac{1}{2}$ ft. in British engines.
In boiler barrels up to about 12 ft. long, two plates
are generally used instead of three plates as in
former practice. For longer barrels three plates
are usual, though two are sometimes used.
Engines have recently been built for the Great
Northern Railway, in which the barrel 11 ft. $5\frac{1}{2}$ ins.
long is made of a single plate. Such plates have

8

BOILER SHOP 9

the advantage of dispensing with a transverse riveted seam, and are consequently stronger, but large plates above certain commercial sizes are more expensive per ton weight.

The plates have to be joined together by riveted joints. Each plate is bent round to a true circle, and its edges which come together are covered with narrow plates both outside and inside the boiler. These narrow plates $A A$ (Fig. 2) are cal ed butt strips, and when fixed properly in place are riveted to the main plate to form a joint running longitudinally along the boiler. This forms one ring of the boiler barrel. The second ring is made similarly, and the two rings have then to be united by rivets to form the complete barrel. The latter joint is termed the transverse joint, and may be made from two rings of the same diameter, in which case a narrow weldless ring B must be riveted all round the circumference over this joint. More frequently, however, the front ring is made of smaller diameter than the back ring, just sufficient to allow it to enter telescopically inside the back ring for a short distance, the two then being riveted directly to each other.

The boiler illustrated in Fig. 2 differs from the one mentioned above in that it has three rings of the same diameter, which are united by single-riveted circular butt strips B, i.e., there is only one row of rivets on each side of the transverse joints. The longitudinal joints A where the ends of each ring are joined together have

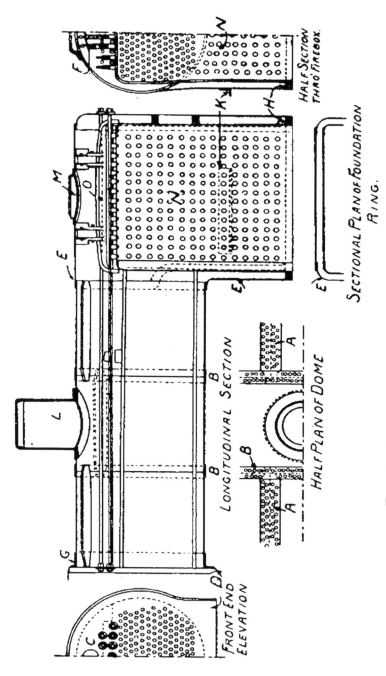

Fig. 2.—Sections through Locomotive Boiler.

BOILER SHOP

double-riveted butt straps, i.e., there are two rows of rivets on each side of the joint.

Before the barrel rings are made and jointed, the straightened plates are marked off for the rivet holes. In British practice the holes are usually drilled, but in some works, and invariably in American practice, they are punched. Punching boiler plates is injurious to the metal immediately surrounding the holes, and therefore such holes are punched smaller than the finished sizes and afterwards enlarged by passing through them a tool in the form of a round drill with cutting edges, known as a "reamer." This operation removes the damaged metal.

The exact position of each rivet-hole has to be marked off on each plate when punching is adopted, but when the holes are drilled five or six plates are taken at a time and only the top plate is marked off. The plates are then secured under a radial drilling machine (Fig. 3) and each hole marked on the top plate is drilled through all the plates by a single operation. The top plate is again used as a template or guide for drilling the holes in the next batch of plates.

The plates have also to be marked off at the edges all round so that they can be finished to the exact sizes. They may be done from the drawings in case only one or two boilers are being made, but when a large number are in hand the marking-off is usually done by template. The template is a sheet of metal cut out like a dressmaker's pattern to the exact size of the

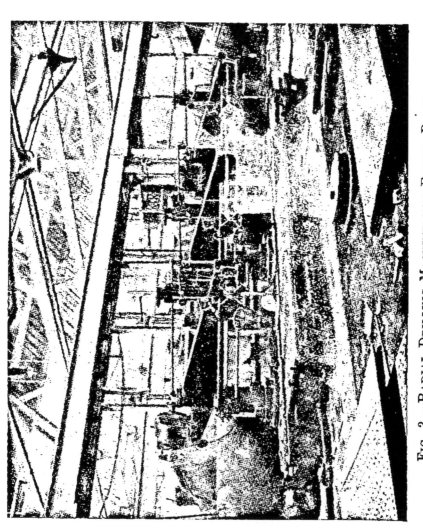

Fig. 3.—Radial Drilling Machine for Boiler Plates.
By Messrs. William Asquith, Ltd. (Halifax).

plate, and in most cases it has also the rivet and other holes drilled through it to show their positions. Finally the edges of the plates are machined to size in a special plate-edge planing machine, in which a cutting tool moves along the edge for the full length or width of the plate.

The next operation is to bend the barrel plates into circular rings. This is done in another form of "mangle," or *plate-bending rolls*. In this machine there are two bottom rolls (Fig. 4) which rotate in fixed bearings, and one top roll, the bearings of which can be moved up and down vertically. By gradually bringing the top roll down each time the plate is passed backwards and forwards through the rolls, a pressure is brought to bear on the plate, and by passing it through a sufficient number of times it is gradually bent to a true circle, except at the ends where the curvature is completed by pressing or hammering it into a block. This last operation is of great importance, and has to be done because it is impossible to finish the edges of the plates to a true circle in the rolls, and unless a true circular form be obtained, the boiler will strain at the joint, and give endless trouble in service.

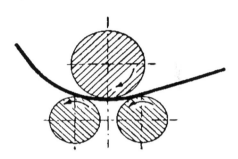

FIG. 4.—DIAGRAM OF PLATE-BENDING ROLLS, IN CROSS SECTION.

FIG. 5.—BOILER SHELL DRILLING MACHINE.
By Messrs. Campbells and Hunter, Ltd. (Leeds).

BOILER SHOP 15

In many works the bulk of the rivet-holes are not drilled until the plates have been bent and put together, only a few holes being made through which bolts are passed to hold the plates together in place. The boiler is afterwards raised on end and secured on the revolving table of a boiler shell drilling machine (Fig. 5), and all the rivet-holes drilled through the plates in position. This produces holes which come accurately opposite each other.

Boiler Tube and Back Plates, Firebox Casing. The *smokebox tubeplate* shown at *C* on the left-hand side of Fig. 2, is a flat plate circular in shape except at the bottom, where there is a projection *D* to fasten it to the cylinders. Its edge is flanged out at right angles all round except along the base of the projection mentioned. The operation of flanging requires considerable skill and great care. It consists in heating the plate in a large furnace to a good red heat, and then sqeezing it between shaped cast iron blocks in a hydraulic press of the type shown in Fig. 6 capable of developing a total pressure of 300 to 400 tons. The table of the press slides up and down on four columns, being actuated by the plungers of the hydraulic rams seen underneath. Heavy cast iron blocks or dies in halves are machined to the shape of the finished plate and jointed together, the corners inside the flanges being rounded to the proper radii. The lower blocks are bolted to the bottom table of the press, and the upper

Fig. 6.—Hydraulic Flanging Press.
By Messrs Fielding & Platt, Ltd. (Gloucester).

BOILER SHOP 17

blocks to the cross-head at the top. The curves and shape of the blocks are so made that when they are pressed together, there is just sufficient space between them for the thickness of the plate. There are two sets of rams, the larger or main rams, and the smaller or vice rams. When the blocks or dies are fixed and the plate to be flanged has been placed in the proper position (in which it is held by steady pins which enter holes in the plate), the vice rams move the table up, and grip the plate firmly between the bottom and top blocks. Then the pressure is put on to the main rams, which squeeze the plate into the required shape between the dies. The whole operation is done at one heat. The firebox back plate is flanged in a similar manner, the dies in this instance being shaped so that the firehole is also pressed to the necessary form.

The " *throat plate* " *E E* (Fig. 2), which connects the firebox shell to the barrel of the boiler, is of a complicated shape, and has to be flanged in two opposite directions, since the upper portion has to have its semicircular flange turned forward to join the boiler barrel, and the bottom part is flanged backwards to connect to the firebox shell wrapper plate. This plate requires extreme care when flanging to prevent damage to the material.

The sides and top *F* of the firebox shell are frequently made in one piece called the *wrapper plate*, which requires no flanging. The top portion is usually circular, and forms a

18   STEAM LOCOMOTIVE CONSTRUCTION

continuation of the boiler barrel. This circular portion is produced by means of the plate bending rolls.

Assembling the Boiler. The boiler barrel is attached to the smokebox tube-plate by means of a continuous weldless angle iron ring G (Fig. 2). This is faced and bored in a large lathe or on a boring mill to fit over the end of the circular barrel plate, on to which it is shrunk by being heated slightly so that it expands. When put into place it contracts and grips the barrel plate, to which it is riveted all round the circumference.

The foundation ring H at the bottom of the firebox (Fig. 2) is a steel casting or forging of rectangular shape, the sides of which must be perfectly " square," as it has to fit accurately between the inner copper firebox and the firebox outer shell or casing. It is machined and ground to accurate dimensions, the rivet-holes marked off and drilled, and then it is fitted to the firebox shell.

The whole of the plates and parts are then put together, and temporarily held in place by bolts through some of the rivet-holes. The structure has to be levelled carefully, so that the different plates are " square " and in line with one another. Careful measurements are made to see that the parts come together in accordance with the drawings or templates, and straightedges are used to test the alignment and " squareness." When this is correct the riveting is done, except

BOILER SHOP 19

the foundation ring, which cannot be riveted until the inner copper firebox is put into place.

Riveting in former years was done entirely by hand, three men and a boy being employed. The boy attended to the heating of the rivets in a suitable portable furnace. One man placed the white hot rivet in the hole, and then held up the head tightly against the plate with a heavy " dolly " tool, which acted as an anvil to take the force of the blows. The other two men on the opposite side then hammered down the protruding shank of the rivet, until a head was formed. In present day practice hydraulic riveting machines are used invariably. These are shaped after the manner of the pincer claws of a lobster, which grip the rivet from both sides of the plates and squeeze it up in the hole. Before any riveting is done it must be seen that the holes come fair and true with each other, and if the marking off has been done properly to template this is usually the case. Any holes which may not come quite true are drilled or reamered out to a slightly larger size, and larger rivets are used in them. Holes which have been originally drilled instead of punched are more accurate in respect to true alignment.

An important point in putting the boiler together is to see that the expansion brackets or angle irons K (Fig. 2), which support the boiler on the frames at the firebox end, are perfectly level, otherwise the boiler may bind on the frame, and not expand properly when hot.

20   STEAM LOCOMOTIVE CONSTRUCTION

Dome and Safety Valve Seating. The dome L (Fig. 2), is formed of a small plate bent into a ring, the joint being either riveted or welded. It is flanged at the bottom in a press by means of blocks made to suitable shape. The flanged portion fits on to the circular barrel plate of the boiler, in which a circular hole has been made, and is riveted to it. The dome cover is a dished plate which is pressed into shape between blocks in a press, or sometimes under a steam hammer. The cover is secured to the dome, which has a faced flange or angle iron ring for the purpose, studs with nuts being used so that the cover can be removed when required. The safety valve seating and cover M are made in a similar manner.

After the boiler has been riveted up, the seams or joints are *caulked*. This is a process for making the joints thoroughly steamtight and preventing leakage, and consists in burring up the edges of the plates with a tool in the form of a chisel with a broad blunt edge. Care must be taken not to use a thin or sharp edge on the chisel or the plates will be injured and their edges forced apart. Caulking which was formerly done by hand, is now done by pneumatic hammers.

Inside Firebox and Stays. The inside firebox N (Fig. 2) is of copper plates. These are planed at the edges, and the rivet and stay holes are marked off and drilled in the same manner as in the steel plates of the boiler and firebox shell. The

BOILER SHOP

flanging, however, is done usually by hand, the plates being heated and hammered down over shaped blocks with wooden mallets. The firehole, the circumference of which has to be flanged outwards to join the doorplate of the firebox shell, is however dished in a hydraulic press. (In Fig. 2 the older form of firehole ring is used instead of the flanged construction.) The flanges are then cut level by placing the plates on the table of a large horizontal band saw, and setting them level for the saw to cut off the ragged edges and leave the flanges of the correct width.

The holes for the tubes in the copper firebox tubeplate, and also in the steel smokebox tubeplate, are marked off either by drawing them out in detail on the plates when an isolated engine is being made, or by marking their positions from a template when several engines are being built.

The firebox plates are fitted to the foundation ring, and are put together in a similar manner to those of the firebox shell. Rivets of very soft iron or steel are generally used for copper fireboxes, though copper rivets have been used in this country and are still used on some French railways. The riveting is done with a hydraulic riveter having a very long gap or jaws, but the pressure used on copper plates is considerably less than that used on the steel plates of the boiler. If the roof stays are of the girder or bar pattern (as at O, Fig. 2), they have to be marked off and fitted to the firebox before the

22   STEAM LOCOMOTIVE CONSTRUCTION

latter is put into and secured to the steel firebox shell.

To put the copper firebox into place the boiler is lifted by a crane and turned upside down, so that it rests on its back supported on blocks with the bottom of the firebox casing upwards. The copper firebox is then lifted by the crane and gently lowered into the firebox casing. To accomplish this, the foundation ring, which has been placed temporarily in position, will have to be removed. The firebox has to be set by measurement and to the marked centre lines, so that it will occupy an exactly central position. When everything is correct, the firebox and the shell are riveted together through the replaced foundation ring with long rivets, and also at the firehole ring. The roof stay holes in both the firebox and the top of the shell, when direct roof stay bolts are used, are reamered out with long reamers, and the holes are tapped in position for the stay screw threads. The reamering and tapping of screw threads through the plates are done by means of pneumatic drilling machines. These are driven at high speed by compressed air, and are arranged to take drills, reamers, taps, etc., as required. It may here be mentioned that every modern boiler shop is equipped with a complete air-compressing plant, from which pipes are taken to every part of the shop. To a large number of points on these pipes, flexible tubes can be attached to serve pneumatic chipping and caulking hammers, portable drilling machines, etc., so

BOILER SHOP 23

that these tools can be used on any portion of a boiler, in whichever part of the shop it may be standing.

The holes for the short side stays are similarly reamered out and tapped.

The copper side stays are turned and screwed in automatic lathes. The threads have to be made very accurately, since it is essential that they should fit tightly into the " tapped " or screwed holes in the copper plates of the firebox and into the steel plates of the firebox shell. After having been screwed into position, either by hand or by a pneumatic machine, the ends of the stays are cut off to a definite length, which leaves a short length projecting on each side of the plates to be riveted over.

Tubes. The tubes are put in from the smokebox end, the holes in the front tubeplate being, for this purpose, made slightly larger than the outside diameter of the body of the tubes. The tubes themselves are enlarged at the smokebox end, and swaged down to a slightly smaller diameter at the other end where they pass into the firebox tubeplate. At the latter end the tubes project into the firebox for about $\frac{3}{8}$ in. to allow for " beading over " the ends up against the copper tubeplate. They are tightened in the holes by means of a tube expander, a special tool provided with a number of rollers. This is placed inside the tubes, the metal of which is rolled outwards until they are a tight fit in the holes. In the

3—(5350A)

24 STEAM LOCOMOTIVE CONSTRUCTION

case of steel tubes a ridge or beading is also rolled on the tube on the boiler side of the firebox tubeplate. Brass or copper tubes are expanded and secured by driving short pieces of steel tube (called ferrules) tightly into the ends which pass through the tubeplate. At the smokebox end the tubes are merely expanded, neither beading nor ferrules being used.

Mountings. The boiler is then sent to another part of the shop to have the " mountings " put on. This term includes all the brass and other fittings, such as safety valves, water and steam gauges, cocks, etc. In British practice the mountings are not screwed directly into the boiler plates. Steel " pads " or seatings of the same shape as the flanges of the mountings having been previously riveted to the boiler, the flanges are secured to these by studs. The joints are faced and scraped, and either made metal to metal with boiled linseed oil, or a thin sheet of asbestos slightly smeared with red lead is placed between the faces. The nuts are then tightened gradually to make an absolutely steam-tight joint.

Boiler Testing. The boiler before it leaves the shops undergoes two tests. The first is a hydraulic test up to a pressure from 25 to 50 per cent. higher than the pressure at which the boiler works in service. The safety valves are locked, the boiler filled up with water, and the necessary pressure obtained by forcing additional water

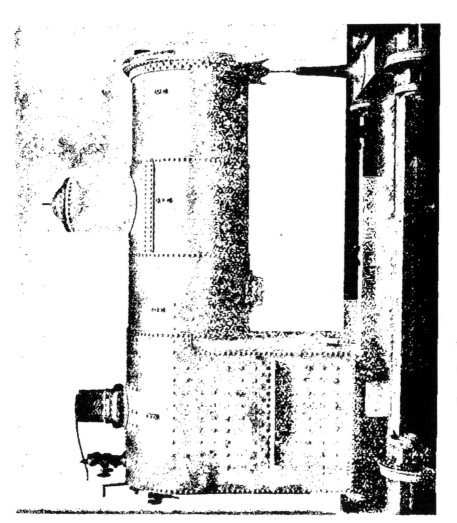

FIG. 7.—COMPLETED LOCOMOTIVE BOILER.

26 STEAM LOCOMOTIVE CONSTRUCTION

in by means of a pump, or by connecting the boiler by a flexible pipe to the hydraulic installation of the shop, suitable pressure valves being inserted in the main to avoid any excess of pressure which would strain the boiler. All leaks and defects have to be made good before the second or steam test is applied.

The steam test pressure is limited to 10 lb. per sq. in. in excess of the working pressure of the boiler, and if this test is passed the boiler is painted with a good coat of anti-corrosive paint to preserve the plates, and is then ready for sending over to the erecting shop to be placed on the engine. Fig. 7 shows a locomotive boiler in its finished condition. *

* Fig. 7 is reproduced from the author's book *The Development of British Locomotive Design,* by kind permission of The Locomotive Publishing Co., Ltd. The opening in the front of the firebox casing is special to this particular boiler, and does not exist in ordinary boilers of standard construction.

CHAPTER III

FOUNDRIES

THE work produced in the iron and brass foundries of a locomotive works is similar to that done in most general engineering foundries, except perhaps that locomotive cylinders are somewhat more complicated than those of most stationary engines. The moulding and casting of metals is a separate trade in itself, of which only the briefest mention can be made here.*

Iron and Steel Castings. The number of locomotive parts made of cast iron is now somewhat less than was formerly the case, steel castings now being used for many of the brackets, hornblocks and other details, on account of the greater strength and reliability of this material. Steel castings are, in fact, replacing many forgings as well. The art of manufacturing steel castings is a special one, and for this reason few locomotive works have their own steel foundries, and most of them find it easier and more economical to purchase steel castings from firms who have made a special study of the subject for years. Steel castings are used for wheel centres, motion

---

* See also *Patternmaking*, by Ben Shaw and James Edgar (Pitman), uniform with this volume. Also *Foundrywork* by the same authors.

27

28 STEAM LOCOMOTIVE CONSTRUCTION

plates, frame stays, horn blocks, driving axle-boxes, dome seatings, bogie centres and various brackets and smaller parts.

Iron castings are used for cylinders, the drag-box under the footplate, the chimney, blast-pipe, and regulator-head in the dome, also for certain horn-blocks, smaller axleboxes and brackets. Brake cylinders, brake blocks, firebars, sandboxes and on some railways the wheel coverings or "splashers" for goods engines are also of cast iron. This list includes castings of very varying types and qualities of cast iron. For instance, the cylinders are made of the very best close-grained iron of special quality and mixtures, and the greatest care is taken to produce a strong hard material free from any "honeycomb" or other defects, but at the same time the hardness must not be so excessive that they cannot be suitably machined and fitted.

At the other end of the list are the firebars, which are usually made of common scrap, much of which has had to be rejected as unsuitable for other foundry purposes.

It must not, however, be understood that scrap iron castings are inferior as a material. All the best castings such as cylinders contain a percentage of scrap iron varying from 15 to 50 per cent., but this scrap is best selected material, derived from old cylinders and best machine parts. The pig irons used are also selected with a view to the quality of the castings desired, and for cylinders it is usual to mix two different

FOUNDRIES 29

good grades of pig iron with a percentage of
scrap. In a locomotive works where a large
variation in the qualities of the different castings
has to be made, the greatest skill on the part of
the foundry foreman is required in selecting and
grading his materials.

Moulding for General Castings. The various
groups of castings require different methods of
moulding. Many of the smaller pieces, of which
a large quantity are required, are moulded by
machine,- or " plate moulded." In the latter
case the patterns are of metal and cast or fixed
on to a plate. If the shape of the article to be
made allows of it, the whole pattern is placed on
one side of the plate, but in many cases half the
pattern is on one side, and the other half on the
opposite side. The mould is then made in two
portions, the second part being produced by
turning the plate over. One of the main objects
of plate moulding (apart from speed of output),
is to save wooden patterns from being damaged
by the constant knocking and shaking required
to remove them from the sand. A large number
of articles can be made from the metal patterns
in plate moulding which are all of the same size
without variation. Vacuum brake cylinders,
which are very thin castings, are made in this way,
and also many forms of smaller axleboxes, brake
blocks, etc.

The large drag-box casting is made by " green
sand " moulding on the floor of the foundry. The

FIG. 8.—AXLEBOX MOULDING MACHINES AND MOULDS.
G.W.R. Swindon Works.

FOUNDRIES 31

bottom half may be moulded in the sand of the floor, and the top half in an open box, which is then placed over the top of the floor mould, and pegged in place, so that the two half-moulds together will produce one casting. Usually, however, two large open boxes are employed, each containing half the mould, and the two are fastened together by pins and cotters. Nearly all the medium sized castings required for a locomotive are moulded in this manner.

Cylinders. Cylinders are the most important and difficult castings made in the foundry. The difficulty lies in the number of complicated passages which have to be moulded by means of sand cores. If a passage has to be made in a casting, that portion of the mould which represents the passage must, of course, be filled up with a piece of special sand made to the shape and dimensions of the passage, so that when the iron is poured into the mould the metal flows round this " core," which remains in place until the casting has cooled. Afterwards, when the sand " core " is removed, the desired passage remains in the casting. These cores are made in wooden core boxes, the two halves of which are made with cavities, and when put together form a hollow pattern for moulding the core desired. The cavity is filled with a special binding sand held together by laces and nails, and the core so formed is then dried in a stove and placed in position in the cylinder mould. To form a support for the

32   STEAM LOCOMOTIVE CONSTRUCTION

core in the mould, the ends of the core are extended to form " core prints," which rest in corresponding recesses in the mould. The steam and exhaust posts are formed from cores of this type, and the hollow cylindrical bore of the cylinder is formed by another large core.

All these cores are made of " loam," composed of clay and sharp clean sand mixed together with cow-hair, which helps to bind the core, but more particularly makes it porous, so that the gases formed when pouring can escape.

Fig. 9 shows a view of the large iron foundry at the Great Western Railway locomotive works at Swindon. The moulds on the floor in the foreground on the left-hand side show single outside cylinders in various stages of the work. Two of these finished cylinders are being lifted by cranes. In the middle of the floor is a half pattern of a pair of inside cylinders, in front of which is an iron core barrel, on which the loam core for the bore of one of the cylinders is plastered, and finally " strickled," i.e., the core barrel with the wet loam on it is rotated and the excess of loam swept off by means of a straight lath of wood until the core is of the proper diameter. After having been dried in a drying chamber, the barrel core is placed in its proper position in the mould. To hold it in position there are " core prints " formed in the mould. The pattern in the middle of the floor shows two of these " prints " on the front face.

All moulds must be rammed and " vented "

FIG. 9.—IRON FOUNDRY.
G.W.R. Swindon Works.

34   STEAM LOCOMOTIVE CONSTRUCTION

carefully to permit the escape of gases formed during the casting of the metal.

Inside cylinders were formerly cast separately and afterwards planed and bolted together to form a pair. Modern practice is to cast both inside cylinders in one piece. This requires more expensive patterns, but saves a lot of machine work.

Brass Foundry All the brass and gun metal castings are made in the brass foundry. These are much smaller pieces than iron castings and consist of axlebox brasses, injectors, cocks and taps, slide valves, and a host of miscellaneous details required in a locomotive. Much of the work is "plate moulding." Various mixtures of metal are used from gun metal bronzes for slide valves and axlebox "brasses" to common brass for taps, etc. The metal is melted, not in large cupolas as in the iron foundry, but in plumbago crucibles which are heated in furnaces generally placed below the floor level.

CHAPTER IV

FORGINGS, SPRINGS, ETC.

Forging and Smithing Departments.—Locomotive forgings include work done in three different departments, to which may be added a fourth, the spring shop. Strictly speaking, the forgings proper are large pieces such as crank and straight axles, connecting rods, etc., which are made under the steam hammers in the large forge. In many cases forged axles are purchased outside from the manufacturers.

A second department consists of the blacksmith's shop, in which smaller parts of wrought iron or mild steel are made by welding. There is also the third or " stamping " department, in which a large number of parts, formerly made by the blacksmith, are now stamped to shape in dies under gravity or " drop " hammers.

All wrought iron and mild steel parts of a locomotive, such as axles, connecting and coupling rods, piston rods, the machinery or " motion," and a large number of other details are made in one or other of these departments. The springs are made in the spring shop by the spring smiths, and their manufacture is an art requiring special knowledge. Frequently they are purchased outside.

35

Axles. As an example of a locomotive forging a simple straight axle may be taken to give an idea of the principles involved. Such axles are made from steel "blooms," which may be described as long thick blocks of steel supplied from the steel works. An axle bloom is about 5 ft. long × 9½ ins. square, the corners being rounded or chamfered off. The end of the heated

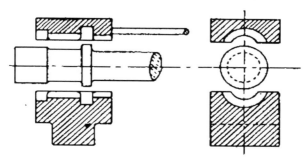

FIG. 10.—BLOCKS FOR FORGING STRAIGHT AXLES.

bloom is placed under the steam hammer and hammered into a circular form. The next operation is to form the journal portion and the collar of the shape shown in Fig. 10. To this end special blocks or dies of steel are machined out to the shape of the forging required. The bottom die fits into the anvil of the steam hammer, the top one being held by the forgeman. The axle is placed on the bottom die, and the hammer brought down upon the back of the top die, and by gradually turning the axle round, it is brought to the required shape. The middle of the axle is circular and is simply hammered

FORGINGS, SPRINGS, ETC. 37

down to shape. After one end of the axle has been finished, the other is treated similarly. As the axle has to be machined, about $\frac{1}{4}$ in. of metal is left all round for turning, and the proper length is tried over with a template.

This method of swaging or forming the shape in dies is characteristic of nearly all locomotive forgings, when any quantity has to be made. Naturally it does not pay to make expensive dies when only one or two similar forgings are required. In such cases the forgeman has to work out the shapes with such ordinary swages and tools as are usually to be found in a forge.

The most difficult forging, and the one requiring the greatest care is that for the crank axle.* The ingot is heated in a large furnace, having at the same time a " porter " bar or staff welded to one end at p (Fig. 11A). The " porter " is a long taper bar, which serves for the handling of the work by the forgeman, and is supported by a crane chain. Frequently it is welded to the ingot whilst the latter is being made. The ingot is first hammered down under a heavy steam hammer to a rectangular shape with chamfered edges as shown in Fig. 11A. This slab will be about 24 or 25 ins. deep × 12 ins. wide, and nearly 6 ft. long, and weighs about $2\frac{3}{4}$ tons without the " porter." Whilst still hot it is taken to a hot saw, close at hand, and four saw cuts are made as shown at a, which extend

* An illustration of a crank axle is given in the primer on *The Steam Railway Locomotive.*

nearly half way through the depth. The part b_1 is then removed either by cutting with a hot saw along the dotted line, or by means of a special cutting tool, and then the remaining portion c (Fig. 11B) is roughly rounded in swage blocks

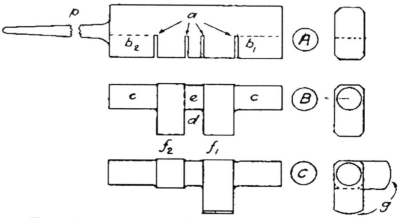

Fig. 11.—Stages in Forging a Crank Axle.

under the steam hammer. The portion d is also removed by means of special cutting tools, and the middle part e between the cranks is roughly swaged down to a circular form. To cut out the portion b_2 the forging is re-heated and the porter bar p is re-welded to the opposite end. After this operation is complete, the appearance of the axle is as in Fig. 11B.

As the two cranks have to be at right angles to each other, one of the two parts $f_1 f_2$ has to be twisted through 90°. This is done by firmly holding the crank f_1, supposing this to be the one to remain in its original position, between the

FORGINGS, SPRINGS, ETC. 39

tup and the anvil of a steam hammer, and then by means of hydraulic or other pressure applied at the end of a very large spanner, to provide sufficient leverage, the other arm f_2 is gradually bent round at a right angle. Lastly the ends of the crank webs are rounded at g under the hammer, and the appearance of the axle is then as shown in Fig. 11c. The right angle must be tested with a square, and all the dimensions checked to gauges to see that enough material is left throughout to enable the machining to be done subsequently to proper finished sizes. The whole of the above operations require several heats, and the greatest care must be taken in this respect, so that no re-heating is done until after mechanical work, e.g., hammering, has been done on the axle, otherwise the internal physical structure of the steel will suffer injury. Finally all crank axles are re-heated and generally cooled in oil, the scientific considerations underlying " heat treatment " being, however, beyond the scope of this primer. Test pieces are taken from the parts machined off the webs of each axle, and subjected to rigorous tensile and cold bending tests. An oil treated crank axle should have an ultimate tensile strength of not less than 35 tons per sq. in. with an elongation of not less than 20 per cent. in a length of 2 ins.

There are other methods of making crank axles by building up the second crank web with slabs welded to the bloom in such a way that the twisting of this web through a right angle is

4—(5350A)

avoided. In some foreign works the webs are swaged in dies in their proper positions at right angles.

Miscellaneous Forgings. Connecting and coupling rods are forged solid out of slabs of mild steel, the middle portions being hammered down to a rectangular section, and the ends, which have to be of special shapes, being swaged in blocks cut out to suit those shapes. The whole work is done under the steam hammer.

Buffer heads form a good example of forging in-die blocks under a steam hammer. They are usually made of wrought iron scrap, piled and heated in the furnace and hammered into " blooms." Each bloom is then reheated and hammered in suitably shaped blocks into the form shown in Fig. 12A, with a shank about 2 ft. long having a ball about 7 in. in diameter at one end. The shank is then placed in other blocks of which a section is shown at *B*, for the steam hammer to flatten the ball and form the head of the buffer. The bottom of the three blocks is firmly secured to the anvil of the steam hammer.

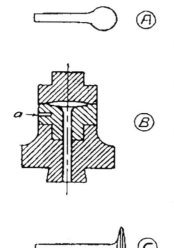

FIG. 12.—BUFFER FORGING.

FORGINGS, SPRINGS, ETC. 41

The middle block is loosely dropped into a recess machined in the bottom block, and has a cup shaped depression at the top to suit the shape of the buffer head when finished. The top block is secured by wedges to the tup of the steam hammer. The two bottom blocks are made with a central hole to take the shank of the buffer. After the buffer head has been hammered to shape the " fin " of metal round it is cut off, after which the finished piece appears as shown at C. The middle block being loose can be rotated by the forgeman during the hammering, by placing a bar in the recess a shown in the side. All such blocks are machined and fitted together, and the exact shape of the forging is cut out in them.

Hydraulic Forging. A considerable amount of forging is now done by hydraulic press, instead of under a steam hammer. In this case the white hot pieces are welded by squeezing them together under pressure, the process being done in blocks or dies as in the case of hammer forgings. At one large railway works some of the casings in which the buffer heads work are made in this way, more especially for wagons. Formerly these used to be of cast iron, but they are now much stronger forgings. Such a finished buffer casing is shown in Fig. 13, and the three portions from which it is forged in Fig. 14. These consist of a flat steel plate C, in which a hole is punched, a conical cylinder B which is bent round a vertical roll and has an open seam a, and a ring A having a diagonal seam which is welded. The ring *A* is pressed on to the cone *B* at a welding heat, and then the combination is again heated and pressed into the plate *C*. The whole of this work is done in suitably shaped dies, so that after the final weld the piece comes out as in Fig. 13.

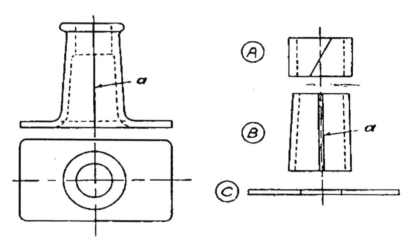

FIG. 13.—FORGED BUFFER CASING.

FIG. 14.—PIECES FOR FORGING BUFFER CASING.

The joint *a* is merely closed together and not welded. The whole of the work is done under a 100 ton hydraulic forging press with two rams horizontal and vertical. One ram holds the work and dies in place, and then the other ram is set in action and squeezes the parts together.

Blacksmith's Shop. In this department the work consists principally in welding rods, bars,

FORGINGS, SPRINGS, ETC. 43

angles, etc., and in the manufacture of brake shafts and levers, hooks, links, couplings, etc. Although this shop is of great importance, it has of recent years lost much of its work because

FIG. 15.—STAMPING SHOP WITH DROP HAMMERS.
By Brett's Patent Lifter, Co., Ltd. (Coventry).

important details of the motion work of locomotives are now made as solid stampings in the drop hammer department instead of being welded. All the smaller parts of the valve gear, etc., are now made as stampings without welds. Steel castings have also superseded much of the wrought iron work formerly done by the blacksmiths.

Stamping Shop. Stampings are made in dies under drop hammers. In these hammers the tup or weight is raised by ropes connected to a rotary steam cylinder placed at the top of the hammer girders, and falls by gravity on to the

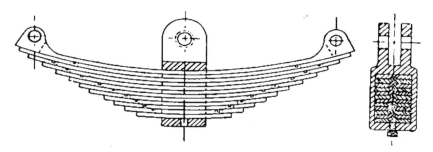

FIG. 16.—UNDERHUNG LAMINATED SPRING.

work, which is placed on the anvil. The pieces made under these hammers are generally much smaller than the forgings produced under steam hammers, and consist of various brackets, links, pistons, parts of valve motion, spring links and hangers, hooks, bell crank levers, hand wheels, spanners, etc. Drop hammer stampings are much cleaner and nearer to shape and size than blacksmiths' forgings, and less work is required in machining and finishing them in the machine and fitters' shops. Fig. 15 shows the stamping

FORGINGS, SPRINGS, ETC. 45

shop with drop hammers in a large locomotive works. A number of drop-forged pieces are shown in the foreground.

Spring Shop. In this department, the springs are made and tempered, the eyes or hooks being welded on to the top plates. Fig. 16 shows a locomotive spring. The plates are maintained in position by small projections or " nibs " on each plate which work in slots cut in the plate immediately below it. The plates have to be " set " to the required curvature, and are then tempered by oil treatment. Finally they are put together and held securely whilst the central hoop or " buckle " is shrunk on. The buckles are forged solid, and machined inside where they pass over the plates. A double eye to take the axlebox pin, is forged on for underhung springs.

CHAPTER V

MACHINE SHOPS—FRAMES AND CYLINDERS

Frames.—The frames come into the frame shop from the steel plate manufacturers, in whose works they are rolled in a similar manner to boiler plates. They are thicker than the latter, the usual thickness in British practice being from 1 in. to $1\frac{1}{8}$ in. They arrive cut to a general all round size as shown by the outer rectangle in

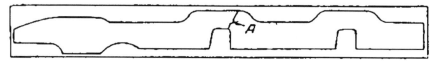

Fig. 17.—Frame Plate.

Fig. 17, this size being the smallest possible to allow of the finished frame being cut out from it. The frame plates are first tried over for straightness, and then levelled and straightened by means of hydraulic jacks.

They are then marked off, and roughly punched out to shape by making a series of interlacing consecutive holes outside the marked lines as in Fig. 18, leaving about $\frac{1}{4}$ in to enable them to be machined to size. After being punched, the plates are heated in a furnace and annealed or cooled slowly to remove injury to the material caused by the punch. When cold they are slotted

46

MACHINE SHOPS—FRAMES AND CYLINDERS 47

to the exact size and shape required in a large slotting machine. All the machines in the frame shop are very large, the bed of the slotting machine being at least 50 ft. long, and the machine has three or four tools working at once. Eight or ten frames are placed one on top of the other,

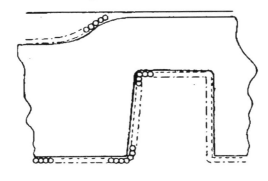

ENLARGED DETAIL OF HOLES
FIG. 18.—PUNCHING THE OUTLINE OF THE FRAME PLATE.

and all are slotted together at one operation, the top frame only having been marked off, and the proper outline drawn upon it. The next operation is to mark off and drill all the bolt and rivet holes, of which there will be from 300 to 400 according to the size and type of engine. The holes for bolts and cold rivets are drilled $\frac{1}{16}$ in. less than the finished diameter, so that they can be reamered or "rose-bitted" out by the erectors when fitting up the cylinders, stays, and brackets to them. About four or five frames are drilled together in one batch, of which the top one is

48 STEAM LOCOMOTIVE CONSTRUCTION

marked off, and this plate is afterwards used as a template or drilling " jig " when the later batches are drilled. In some railway works separate frame templates or jigs are used for standard classes of engines. These consist of plates about $\frac{3}{4}$ in. thick in which all the holes are drilled in their exact positions but larger than the holes in the frame. The holes in the template or " jig " are then bushed with steel bushes, the internal diameters of which are the exact diameters of the holes as drilled in the frame. This jig is placed on the top of the frames to be drilled, and the bushes act as guides for the drills. The drilling is done in multiple radial drilling machines, which may be either special machines or of the radial type shown in Fig. 3.

The drilling completed, the frames are placed upside down in a vertical position in screw jacks or trestles provided with arrangements for holding them in this position. All sharp edges are then removed and the corners of the recesses into which the hornblocks fit are rounded. The frames are again laid on their sides, the hornblocks are bedded down and fitted into the recesses slotted in the frames to receive them, and the various brackets are also bedded down to their proper places. The bolts which secure the hornblocks are turned to exact size, and must be driven in to be an absolutely tight fit. The main frames are then sent to the erecting shop, a pair being required for each engine. The bogie requires an additional pair of small frames. These together

MACHINE SHOPS—FRAMES AND CYLINDERS

with the buffer beams of steel, cross-stays, etc., are dealt with in the frame shop in a similar manner to the frames themselves. The riveting together of the cross-stays and frames is always done in the erecting shop, only the plates themselves being treated in the frame shop.

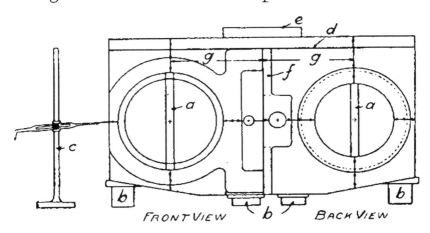

FIG. 19.—MARKING-OFF CYLINDERS.

Cylinders. The cylinders are machined in a shop which contains a number of single and double boring machines, and also drilling and planing machines, all being tools of large size and most of them of special designs for locomotive work. The rough cylinder castings are lifted by a crane on to a large cast iron marking-off table, the surface of which is perfectly level. The casting is placed on planed blocks b, Fig. 19, and its level is adjusted by wedges above them. Those parts on which lines are to be marked are painted with whiting to show the scriber marks, which when made are marked with centre punch dots. Centre strips *a* of thin iron are placed diametrically across the bores. The centres at both ends are found with compasses, and marked on the strips. The cylinders are then levelled by using a straight edge and spirit level across the centres, the necessary adjustments of level being made by the wedges. The centres are tested by the scribing block *c* which has the underside of its foot a perfectly plane surface; the needle pointer of this is shown on the centre line, which is drawn by it all round the casting. All lines such as *d* and *e* for surfaces which require planing are marked. Vertical centre and other vertical lines are drawn with the blade of a T-square having its broad edge on the table. The bore of the cylinder and all holes are drawn with compasses and centre popped. The dimensions are taken from the drawings, or in the case of standard engines will be measured from gauges specially made for the cylinders of different classes of engines. The whole work must be checked carefully to see that there is sufficient material on all surfaces for planing and boring to finished sizes.

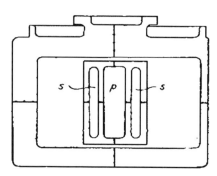

Fig. 20.—Marking-Off Cylinder Port Faces.

The face into which the steam ports (*s*) and

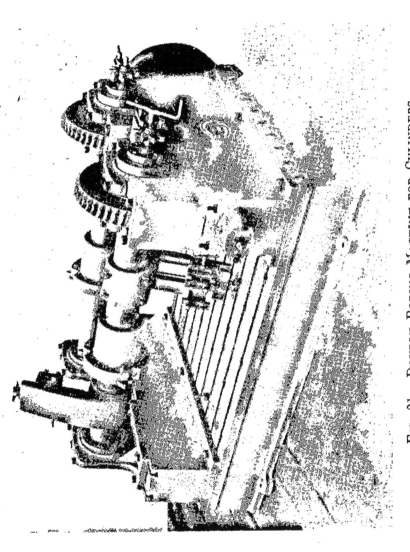

FIG. 21.—DOUBLE BORING MACHINE FOR CYLINDERS
By Craven Bros., Ltd. (Manchester).

52 STEAM LOCOMOTIVE CONSTRUCTION

exhaust ports (*p*) Fig. 20, emerge must also be marked off, a template in which holes are cut for the ports being used for this purpose. The edges of the ports were formerly slotted out, but they are now generally milled out by a milling machine. It should be added that the cylinders shown in Fig. 19 are cast separately, the two halves being bolted together to form one pair. One of the flanges is shown at *f*, Fig. 19, and in this case these flanges have also to be marked off for planing, the essential point being that the dimensions *g*, *g* from the face of the finished flange to the centre of the cylinder on each side are exactly equal.

In many works much of the planing of the various outer surfaces is done first, using the scribed lines as a guide, and then the cylinders are set down on the bed of a boring machine on one of the planed surfaces and bored. The centre line of the bore of the cylinder is the datum line to which all other dimensions refer, and therefore some works consider it preferable to bore the cylinders before planing them. Fig. 21 shows a double boring machine for the purpose of boring two inside cylinders at once. The two heavy revolving boring bars are passed through the cylinders, the latter being bolted to the grooves in the bed of the machine. On each bar there is a disc carrying three cutting tools, and the two projecting arms carry tools for facing the ends of the cylinders, on to which the covers have to be fitted. Each cylinder is

FIG. 22.—CYLINDER BORING AND DRILLING MACHINE. By Messrs. William Asquith, Ltd. (Halifax).

54   STEAM LOCOMOTIVE CONSTRUCTION

bored to exact diameter, except at the two ends where the bore is slightly enlarged. These enlarged ends are termed the " counter-bores " and their object is to prevent a ridge being formed at the ends of the cylinder as the piston wears down the metal. The end flanges for the cylinder covers are also faced at the same time by the tools carried in the projecting arms. These flanged faces and the bores and counter bores can be seen in the pair of cylinders shown in Fig. 22. These cylinders have circular piston valves, and the boring of these is also done on a similar machine with smaller boring bars.

The planing of the various sides and flanges, and also of the steam chests port facings, when the latter have ordinary slide valves, may be done either on an ordinary large planing machine or on special machines made for the purpose. The cylinders have to be set on the machine by the bore to plane the top and bottom surfaces.

In those works in which the boring is done before the planing, one method is to use two standards which are splined underneath, the tongue or spline fitting into one of the grooves in the table of the planing machine. The top part of each standard is provided with a circular face plate which just fits into the counterbore of the bored cylinder. The standards are placed so that the face plates enter the counterbore at each end, and they are pulled up to grip the cylinders by means of a long central bar with nuts. The centre line of the bore is thus held

MACHINE SHOPS—FRAMES AND CYLINDERS 55

perfectly parallel to the bed of the table so that all surfaces parallel to this can be planed accurately.

A large number of holes have to be drilled in the flanges for the cylinder and steam chest covers, and for uniting the cylinders to the frames. Fig. 22 illustrates a modern machine by Messrs. William Asquith, Halifax, for doing this work. It is also used for facing the front flanges to which the cylinder and steam chest covers are secured.

All joints are carefully faced, and covers fitted on to be perfectly steam-tight. The cylinders should then be tested, preferably by steam pressure, though hydraulic pressure is sometimes used.

CHAPTER VI

MACHINE SHOPS—AXLES AND WHEELS

Axles. Straight axles are turned to gauges in special axle lathes. Formerly these lathes were of the ordinary form with a headstock at the left-hand end and a poppet head or tailstock at the right. The most modern form of lathe is of the centre-driven type, in which the axle—after having been turned in the middle—is gripped

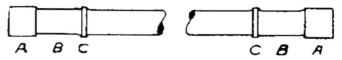

Fig. 23.—Straight Axle.

and driven there, supporting tail-stocks being used at each end. This method gives a more rigid support for the axle, and enables cutting tools to be used at both ends. Fig. 24 shows a lathe of this type, which admits axles 9 ft. long. A straight axle is illustrated in Fig. 23, in which AA are the "wheel seats," i.e., the portions on to which the wheel centres are pressed, and BB are the "journals" which revolve in the bearings in the axle-boxes. The collars CC keep the journals from moving laterally. The wheel seats and journals are turned smooth and truly circular and the journals are finished with

Fig. 24.—Centre Driven Axle Lathe.
By Messrs. Hulse & Co., Ltd. (Manchester).

58   STEAM LOCOMOTIVE CONSTRUCTION

burnishing rollers. These are revolving steel rollers secured in the tool posts, or carried by attachments at the back of the lathe (as in Fig. 24) ; they are pressed against the revolving journals, and impart to the latter a perfectly even and polished surface. Grinding machinery is now frequently used for axles, more especially for finishing the wheel seats.

Crank axles are turned in heavy lathes of the ordinary pattern, as the position of the crank webs does not allow of centre driving. The inside faces of the jaws of the crank webs-are slotted out, or more usually milled out by a large revolving disc, into the periphery of which a number of cutters are inserted. A machine of this type is shown in Fig. 25. This has two centring headstocks at the back (one of which is hidden from view), in which the crank axle is held in special chucks, and there is also the revolving disc provided with cutters. The latter is slowly fed into the crank between the crank arms, and cuts these out to the dimensions required. This part of the work completed, the headstocks are set in motion to revolve the crank axle, which is set so that the crank pin is rounded by a cutting tool to a circular form. The axle is then turned round end to end in the centring headstocks and the other crank arms are milled out and the pin turned.

Many crank axles of large modern engines, instead of being forged in one piece, are now built up of nine pieces, two ends and one middle

Fig. 25.—Crank Sweep Milling Machine
By Messrs. Craven Bros. (Manchester) Ltd.

60 STEAM LOCOMOTIVE CONSTRUCTION

piece, two pieces for the crank pins, all the above being of circular form, and four flat slabs for the webs. The slabs are milled round in a vertical milling machine, the holes for the pins and parts of the axle bored. They are then heated and shrunk on to the circular parts and crank pins of the axle, and finally secured by round plugs screwed into the joints in a direction parallel to the length of the axle.

Wheel Centres. The term " wheel centre " refers to that portion of the wheel which includes the central boss, and the spokes and rim, but does not include the axle or tyre. Wheel centres are now generally made of cast steel. They are turned outside, and also bored for the axles either in a wheel lathe, or—more conveniently—on the revolving table of a horizontal boring mill of the type shown in Fig. 26.

The wheel centres are forced on to the axles in a hydraulic press, the pressure required being from 60 to 120 tons, according to the size of wheel and axle. Small bogie and tender wheels are pressed on by 60 tons or so, but large driving wheels require from 90 to 120 tons. To obtain the necessary grip, the part of the axle or "wheel seat," on which the wheel fits, is turned slightly larger than the bore of the wheel centre. The amount of this difference or " allowance " for an axle $8\frac{3}{4}$ in. diameter is about $\frac{21}{1000}$ in., the hydraulic pressure then required being 10 to 12 tons per inch diameter of the axle. If the pressure

FIG. 26.—HORIZONTAL TURNING AND BORING MILL.
By Messrs. Craven Bros. (Manchester) Ltd.

registered on the gauge or recorder attached to the press is too small the wheel is taken off again, and another substituted. In the case of driving and coupled wheels, additional security is provided by driving keys into keyways cut into both wheel and axle, but in the case of other wheels no keys are used.

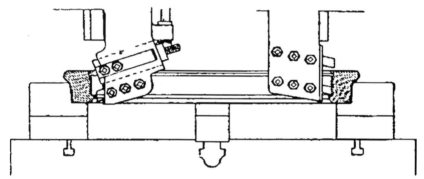

Fig. 27.—Boring Tyres on Horizontal Boring Mill.

Tyres. The tyres, which come from the steel manufacturers, are rolled without weld. They are bored inside to an internal diameter slightly less than the outside diameter of the wheel centre, on to which they have to be shrunk, the allowance being about $\frac{1}{1000}$ of the diameter of the wheel centre. Formerly tyres were bored in a wheel lathe, but it is found much more convenient and expeditious to bore them on a boring and turning mill of the type shown in Fig. 26. The tyres are placed horizontally on a revolving table as shown in Fig. 27. The tool holders

MACHINE SHOPS—AXLES AND WHEELS 63

shown are special ones which allow several tools to be used at once.

After having been bored the tyres are placed on the floor inside a shallow pit, which has a circular gas ring round it, by which they are heated so that they expand. The wheel centre on the axle is then taken up by a crane and lowered into the tyre, which as it cools, shrinks and grips the wheel.

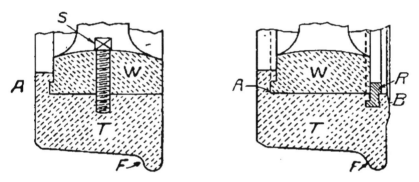

FIGS. 28 AND 29.—TYRE FASTENINGS.

Figs. 28 and 29 show the most usual methods of securing the tyres to the wheels. T is a section through the tyre, and W one through the rim of the wheel. In both examples, there is a lip A which prevents the tyre from sliding laterally over the wheel rim towards the inside of the rail. The flange F, which runs against the rail, prevents the tyre from sliding towards the outside. In Fig. 28 the tyre is secured by set screws S which pass through the rim of the wheel. Some engineers object to set screws, as the tyres tend to

64   STEAM LOCOMOTIVE CONSTRUCTION

break across the holes. In the alternative construction, Fig. 29, the tyre is held by a retaining ring *R*, which is heated and driven into the recessed groove shown, and the lip *B* of the tyre is then hammered down all round the wheel on to the ring.

The outside crank pins, upon which work the coupling rods, and also the connecting rods of outside cylinder engines, are turned and then ground on a grinding machine, and pressed into the holes bored into the wheel boss to receive them. As the pin on one side of the engine has to be fixed at exactly 90° with the pin on the opposite side, the boring of the holes in the wheels is done on a special " quartering machine," in which both holes are bored simultaneously in the opposite wheels. The arrangement of the machine is such that the 90° angle is automatically maintained.

Finally the complete set of wheels and axle are lifted into a heavy wheel lathe, of the type shown in Fig. 30, to have the tyres turned on the outside. These lathes are very rigid and powerful machines, since the extremely hard steel of the tyres places a heavy stress on both the cutting tools and the machine. Figs. 31 to 34 show a set of sheet iron gauges used for turning the tyres. Figs. 31 to 33 are used to see that the cone of the head of the tyre (usually 1 in 20) and the height of the flange are correct. Fig. 34 tests the proper distance apart or the " gauge " of the two tyres on a pair of wheels.

Fig. 30.—Locomotive Driving Wheel Lathe.
By Messrs. Craven Bros. (Manchester) Ltd.

Wheel Balancing. In some shops the wheels when finished are balanced by revolving them at high speed in a special machine, and adding small weights until they revolve with perfect steadiness. A machine for this purpose designed by Mr. G. J. Churchward, chief mechanical engineer of the Great Western Railway, and used at Swindon

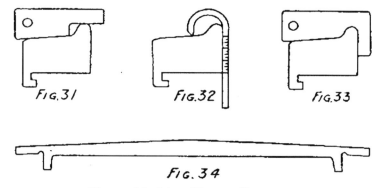

FIGS. 31–34.—TYRE GAUGES.

Works, is illustrated in Fig. 35. The wheels and axle rest in bearings supported by springs, and are driven at a speed corresponding to 60 m.p.h. by means of a 55 h.p. electric motor. The speed is registered by the counter on the wall at the back of the machine. The hand wheel actuates the brake blocks by means of a screw, so that the wheels can be stopped quickly when required after the current has been cut off. The excess or deficiency of balance, as found by the small weights attached during the test to secure steady running, is afterwards rectified in the balance weights.

Axleboxes. The account of the work done in the wheel shop may conveniently include the axleboxes. These are most frequently machined in the general machine shop, but as they have to be fitted to the journals they are, in some modern works, machined in the wheel shop, where they are close to the wheels, and do not then require to be taken from one shop to another. An alternative method, used in other works, is to fit and bed the axleboxes on to a short "dummy" axle journal, kept in the machine shop for this purpose, the axleboxes then being taken straight to the erecting shop.

Axleboxes are either of gun-metal or of cast steel fitted with gun-metal bearings, the latter form being now more usual. The "brasses" which bear on the journals are provided with recesses, into which is run "white" or antifriction metal, consisting of about 80 per cent. of tin, 10 per cent. of copper, and 10 per cent. of antimony.* This prevents the journal of the axle from being injured in case the bearing runs hot.

The boxes are arranged on a planing machine, in two rows of about ten boxes in each row, and planed between the jaws to receive the brasses and axlebox keeps, which are machined in turn and fitted into the boxes. The brasses must be fitted very carefully, and they must be well bedded down to avoid any rocking action of the

* The composition of "white metals" varies considerably. The above figures give one mixture used for locomotive bearings. Many white metals contain from 40 to 70 per cent. of lead, which replaces the tin.

Fig. 35.—Wheel Balancing Machine.
G.W.R. Swindon Works.

MACHINE SHOPS—AXLES AND WHEELS 69

brass in the body of the box. The holes are then broached out for the pin, which secures the " keep " in the box. Following this the boxes are fixed to a long planed casting or angle plate on the bed of a planing machine, about twelve boxes, six on each side, and are planed on one face, and along the recess at the side which fits on to the hornblocks. The boxes are then reversed and the other face and recess planed. They are now bored four or six together on a special boring machine. In some works the boring and the facing are done together on a double boring mill, two boxes being machined at once. Oil grooves, which were formerly chipped out with a chisel, are now frequently grooved out on a small milling machine. Finally the fitters take the boxes and bed the brasses down on to the journals or dummies, by using red " marking " and filing down the high spots.

CHAPTER VII

GENERAL MACHINE SHOP

Machinery and Methods—General. To describe the many processes which various parts of the engine undergo in the general machine shop would be a task far beyond the possibilities of this book. This shop is provided with the lathes, planing, shaping, and slotting machines such as are to be found in all large engineering shops, and in addition there are many special modern tools, such as automatic lathes for making stays, pins, studs, etc.; grinding machinery, and above all, milling machines. Formerly the machining was performed on the first mentioned machine tools of standard pattern, and then passed on to the fitters to be finished by filing, scraping and the other well known fitters' methods. Nowadays the work is done and completed as far as possible on milling and grinding machines, so that as little fitting as possible is required, many of the parts going direct from the machines to the erecting shop to be put together on the engine. The former processes of planing, shaping and slotting are now very largely replaced by milling. In milling machines the work is machined by revolving vertical or horizontal serrated cutters, which produce a fine finished smooth surface. These cutters can be shaped, or two or three of

GENERAL MACHINE SHOP 71

them put together, to mill out irregular curves and grooves, and so save a vast amount of labour formerly required from the fitters. Slide bars, piston rods, and many other parts are ground to size on grinding machines by means of emery or corundum wheels revolving at high speeds. The work is all done to limit gauges, and can be finished within 0·0005 in. if required.

When a large quantity of similar pieces are to be made, they are machined in " jigs " or. " fixtures." These are attachments bolted to the machine, and made to suit the particular pieces to be machined. The latter are secured in or by the jig, which locates their position in such a way that the machining is done accurately without preliminary marking-out of holes and contour lines. If only one or two engines of the same class are being built, it does not pay to make expensive jigs and fixtures, and the older methods of marking off and machining are then employed. The object of the modern improvements is to reduce the cost of the work, and at the same time to improve its accuracy and quality

Pistons, and Piston Rods. The pistons are usually of cast iron, but are now frequently made of stamped steel, which has the advantage of enabling a lighter piston to be used owing to the greater strength of the material. The latest practice is to machine the pistons on a heavy turret lathe, in which a number of tools are fixed in a revolving turret. When one operation

6—(5350A)

has been completed, the turret is turned through an angle to present another tool to the piston for the next operation, and the first tool remains in correct adjustment for use on the next piston in due course. The holes for the piston rod are bored with a large twist drill; and the next tool

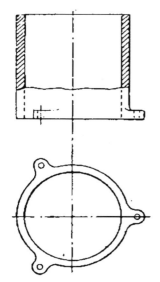
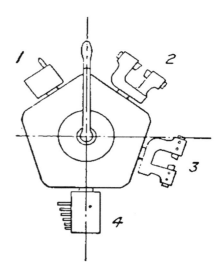

Fig. 36.—Cast Iron Cylinder for Piston Rings.

Fig. 37.—Tools in Turret for Machining Piston Rings.

faces the piston on one side. A third tool turns the outside periphery, and a fourth operation cuts the two grooves in the periphery, into which the piston rings fit. These rings, two in number, are of cast iron, generally about $\frac{1}{2}$ in. to $\frac{5}{8}$ in. broad by $\frac{3}{8}$ in. to $\frac{1}{2}$ in. deep. The outside diameter of the ring is turned about $\frac{1}{8}$ in. to $\frac{1}{4}$ in. larger

GENERAL MACHINE SHOP 73

than the bore of the cylinder. A piece is then
cut out with a saw or parting tool and this enables
the two ends to be pressed together so that the
ring can be sprung into the cylinder. When in
place .the rings spring back, and their pressure
against the cylinder walls causes the piston to be
steam tight. The latest method of making these
rings is to cast a cylinder of the form shown in
Fig. 36 with lugs at the bottom, which can be
bolted to the horizontal table of a turning and
boring mill. This machine is provided with a
turret head which has four tools' (Fig. 37) for four
operations. Tool No. 1 faces the top of the casting,
and is then followed by No. 2 in which two cutting
tools are employed to rough-bore the inside and
rough-turn the outside at one operation. No. 3
is a similar tool for the finishing cuts of the boring
and turning. A depth of 5 or 6 inches of the
casting is now ready to be cut off into rings. This
is done by means of a special holder No. 4 containing six parting tools, so arranged that the
top tool stands out beyond the next one below
it and so on. Each tool enters the casting a little
before the tool below it and thus each ring is cut
off before the one below it, for it is of course
impossible to cut off all the rings simultaneously.

The piston rods are either screwed into the
pistons or have plain coned ends which are
secured to the piston by a large nut screwed
on the front side. The rods are turned nearly
to size in a lathe, and finished to the exact diameter
in a grinding machine. In this machine an

emery wheel *W* (Fig. 38), revolves at a surface speed of about 6,000 ft. per min. The rod *R* also revolves, the directions of revolution being as shown by the arrows. Finally either the rod or the wheel must have a longitudinal motion so that the wheel can grind the whole length of the rod. The direction of this motion is reversed at the end of each traverse. *A* is a clamp for holding the rod in its proper position

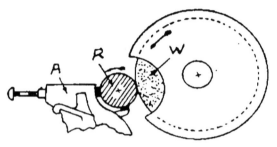

Fig. 38.—Grinding a Piston Rod.

against the wheel. The bed of the machine, not shown, is somewhat similar to that of a lathe.

Crossheads and Slide Bars. Crossheads, which form the attachment of the piston rod to the connecting rod, are of many different forms, and various methods of machining them are used according to their shape. They are planed between the jaws which embrace the connecting rod, four or five being fixed in an inverted position on a planing machine for this purpose. The turning and boring involve several operations,

GENERAL MACHINE SHOP

which may be done in a turret lathe with several sets of tools, each of which in turn does part of the work. In some shops much of this work is done on a horizontal turning and boring mill. Fig. 39 shows a crosshead for two slide bars, from which its complicated form will be seen. The piston rod end fits into the taper hole A; the connecting rod small end fits between the jaws B, and is secured by a pin which passes through the holes C. The slide blocks D, shown in chain dotted lines are separate pieces with channels planed in them to fit over the slide bars.

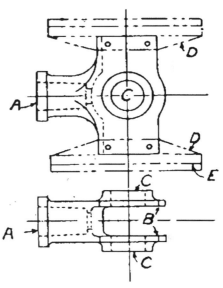

FIG. 39.—CROSSHEAD AND SLIDE BLOCKS.

The slide bars are plain rectangular bars of steel which are planed to within a fraction of the finished sizes, the ends faced, and the holes drilled for the bolts which connect them to the cylinders at one end and to the motion plate at the other. The faces on which the slide blocks move were formerly draw-filed to a true surface on a surface plate. This involved much fitters' work. Now the bars are placed on the reciprocating table of a grinding machine, and the surfaces are ground

STEAM LOCOMOTIVE CONSTRUCTION

true by emery or corundum wheels. Beyond removing the sharp edges no fitters' work is required, and the bars go straight to the erecting shop. The type of grinding machine used is quite different from that employed for grinding piston rods, and may be likened to a planing machine, in which cutting tools are replaced by an abrasive wheel.

Connecting and Coupling Rods. The machining of connecting and coupling rods is a case in which large milling machines have replaced planing machines. The rods are now fluted out to form an I-section, and the whole of the surfaces and edges are milled. Fig. 40 shows the surfaces of two rods being milled on a large machine. In the ordinary type of connecting rod the brasses must be bedded down and fitted into the straps, and the straps fitted to the rods, etc., so that there is no shake, and the rod works as if it were a solid piece. The rod must be straight, and the holes through the brasses, in which the crank and crosshead pins work, must be at right angles to the axis of the rod, and the axes of both holes parallel to each other. Such work has to be done by the fitters when putting the various pieces of the rods together, before sending them to the erecting shop.

Coupling rods in this country are not fitted with adjustable brasses like connecting rods, but have plain circular bushes lined with white metal. On the Great Western Railway the

FIG. 40.—CONNECTING ROD MILLING MACHINE.
By Messrs. T. Butler & Co. (Halifax).

78   STEAM LOCOMOTIVE CONSTRUCTION

connecting rod big-ends are made with similar bushes, which are much simpler, and are stated to wear well for a long period.

Link Motion.—The link motion,* consists of a large number of parts which have to be machined and fitted together. The side faces of the curved quadrant links are planed or.milled to the proper thickness, several being machined at one time. A template is then laid on them and each marked off to the proper finished curved shape. All the holes are drilled through " jigs " which locate them accurately, and the links are slotted out to the lines marked by the template. The blocks which work in the quadrant link are also slotted or milled to shape. The levers which connect the various parts of the link motion are milled on the faces and to contour on the sides. All work of this kind is produced in a better manner and in less time on milling machines than by slotting and planing. All holes and the pins which fit into them are bored and turned to limit gauges. The fitters then fit all the parts together, testing and levelling them so that everything is perfectly " square," and all the necessary oil grooves are cut in the blocks and pin holes.

Case-Hardening. The parts of the link motion are then taken apart and sent to the case-hardening shop, a department which has not so far been

---

* The link motion and also connecting rods are illustrated and described in the primer on *The Steam Railway Locomotive.*

GENERAL MACHINE SHOP 79

mentioned. " Case-hardening " is a carbonizing process which produces a hard steel surface on the parts, without which they would be subject to undue wear. All link motion parts and pins, and ·also slide bars undergo this process. The parts are placed in cases, and packed round with pieces of charcoal, leather, and bones, and carefully sealed up. The cases are then heated in a furnace for 12 to 18 hrs. after which the parts are removed and cooled in water. Case-hardening has been described as a somewhat brutal process, though nothing better has yet been found. It has the disadvantage that the parts so treated warp somewhat in the process, and have therefore to be returned to the fitters and machine shop to be corrected. The case-hardened surfaces are so hard that they cannot be touched with a file, hence the holes are ground out to limit gauges by very small emery wheels or, alternatively, they are "lapped " out by revolving spindles made of lead and covered with emery powder and oil. Due allowance for the warping is made in the previous machining and fitting, a very slight amount of extra material being left for the final grinding to finished size after case-hardening.

CHAPTER VIII

ERECTING SHOP

Frames. In the erecting shop all the parts which make up the complete locomotive are assembled and put together. The frame plates, which form the foundation of the engine, are first taken in hand, and are placed in position on opposite sides of a long pit. This pit is made in the floor of the shop between the rails on which the engine will stand later after the wheels have been placed in position. The frames are placed vertically in their natural position, and rest in grooves made in the heads of screw jacks. By means of these jacks the frames can be levelled, and set in the exact positions required. The frame plates are held temporarily at their proper distance apart by long bolts and distance pieces, until the permanent stays are fastened to them. They are then levelled both longitudinally and transversely by means of long straight edges upon which spirit levels are placed. Their vertical adjustment is made by plumb lines. Both frames must be placed exactly opposite to each other in their proper relative positions. This adjustment is made by placing straight edges across the faces of the hornblocks, and using "trammels." A trammel is a long steel rectangular bar with sliding compass points, the latter being set to a

80

ERECTING SHOP

definite distance apart according to the dimensions on the drawings. The trammelling is done diagonally across the frames between points marked on their top edges immediately above the line representing the centres of the axles. This is shown in Fig. 41 in which FF represent the two frames in plan, and A_1 A_2 A_3 the centre lines of the axles. The dotted lines show the

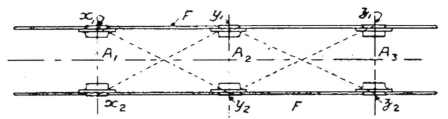

FIG. 41.—TRAMMELLING THE FRAMES.

diagonal directions of the trammels. By trammelling diagonally the "squareness" of the frames is ensured when the distance x_1y_2 is equal to the distance x_2y_1, and y_1z_2 is equal to y_2z_1. At the same time the longitudinal and transverse dimensions must be also checked by trammelling, and x_1y_1 must be equal to x_2y_2 longitudinally, and x_1x_2 equal to y_1y_2 transversely, similar measurements being made for axles A_2 and A_3. The motion plate f (Fig. 42), the cross stay g in front of the firebox, and the drag-box casting h are then temporarily bolted in place to secure the frames together, the bolts used being smaller than the holes in the frames, so that any necessary adjustments can afterwards be made

82 STEAM LOCOMOTIVE CONSTRUCTION

by slightly moving the parts. It is essential that the engine should be built perfectly " square," otherwise it will not run freely, and endless trouble and extra repairs will result.

Fixing the Cylinders. The cylinders are then lowered into place by a crane, and fastened temporarily by ordinary- bolts. They rest upon two screw jacks which are packed up on strong planks placed across the pit. In place of the jacks some shops use angle irons fixed temporarily to the frames as supports for the cylinders. It is absolutely necessary that the centre lines of the cylinders, when the latter are fixed permanently, should be exactly parallel to the frames, and the faces of the hornblocks, which determine the alignment of the axles, must be exactly at right angles to the cylinder centre lines.

A long steel straight edge a (Fig. 42) is fastened across the front of the frames at a suitable height, which may be measured from the drawings. Another straight edge b is fixed across the driving hornblock faces. Thin wires representing the cylinder centre lines are stretched tightly between the two straight edges, their distances apart and from the two frames being measured from the drawings. These wires it will be noticed pass through the bores of the cylinders. In place of the straight edge a, it is very usual to fasten the front ends of the two wires to steel strips fitted across the front end of each cylinder, each strip having a small hole exactly in the centre of the

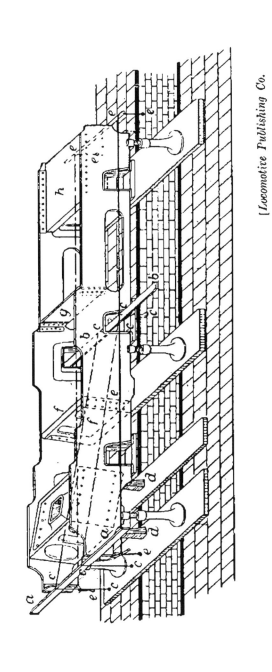

[*Locomotive Publishing Co.*

Fig. 42.—Erecting Locomotive Frames and Cylinders.

84   STEAM LOCOMOTIVE CONSTRUCTION

bore of the cylinder, into which the wire is fastened. The cylinders can be raised or lowered on the jacks and adjusted slightly in a longitudinal and inclined direction until they are in their exact position according to the drawings. Two measurements are necessary to verify this. The circumference of the bore of each cylinder must be equidistant all round from the central wire throughout the full length of the bore. This measurement is tested with callipers. Finally the longitudinal distance of the face of the cylinders from the centre of the driving axle as shown on the straight-edge *b*, is measured on both sides of the engine with a rod of the exact length required. This fixes the longitudinal position of the cylinders.

Two longitudinal wires *e* are also stretched outside each of the frames. These are set with a long square applied to the straight-edge *b*, to which they must be at right angles, and they must also be parallel to the wires *c* representing the cylinder centre lines. The straightness and the parallelism of the frames are tested from these lines.

Finally the exact position of the motion plate *f* is located from the cylinder centre lines *cc*, since the slide bars are fixed to this plate, and it is necessary for the bore of the cylinder, the piston rod, and the slide bars to have the same centre line.

Permanent Fixing of Frame Attachments. So far the parts have only been bolted up temporarily.

ERECTING SHOP 85

Before any permanent fixing is done the foreman of the shop has to go over and test the whole of the work, taking measurements longitudinally, transversely, and diagonally to see that frames, cylinders and cross stays are perfectly square and that all dimensions are correct in accordance with the drawings.

If everything be found in order, all the holes are broached or reamered out to finished sizes, so that where the cylinders or stays are to be united to the frames all the holes in both parts coincide exactly. All bolts and cold rivets connecting these parts to the frames are turned and made a tight driving fit in the holes. The cylinders are bolted to the frames since they may have to be removed for repairs at some future period, and the same applies to the horn-blocks, but the motion plate is cold-riveted. The plate g is usually hot-riveted. Various brackets may now be fastened in position, amongst which are angle brackets placed outside the frames at right angles to them. Long angle irons are riveted to the top of the frames. These angle irons and brackets carry the plates which form the platforms along each side of the engine. Occasionally these platforms are placed above the level of the tops of the wheels, but it is more usual for them to be below this level, in which case there are rectangular holes made in them through which pass the upper portions of the wheels. These holes are covered with segmental casings, known as " splashers," which cover in

86 STEAM LOCOMOTIVE CONSTRUCTION

the protruding portions of the wheels. The platforms, angle irons, and brackets are riveted on by means of portable hydraulic riveting machines similar to those used in the boiler shop.

Erecting the Motion. The cylinders and motion plate being in position the slide bars may now be fixed. For this purpose the wires cc (Fig. 42), representing the cylinder centre lines, are used, and from these the erector can set the bars parallel to the centre lines and equidistant from them above and below. The bars are bolted to faced projections on the back cylinder covers at one end and on the motion plate at the other. Thin brass liners are placed between the bars and these projections, their thickness being adjusted to suit the proper distance apart of the bars. The pistons, with their rings sprung into place and with the piston rods attached, are put in from the front end of the cylinders, and the crossheads are then attached to the piston rods. The crossheads with their slide blocks act as guides in setting the slide bars, between which the slide blocks must work up and down quite freely, but without shake.

The slide valves are placed inside their buckles. These are rectangular frames forged solid with the spindles which pass through the glands in the ends of the steam chest covers. The buckles or frames simply embrace the valves, so that when the spindles are moved by the valve gear, the

ERECTING SHOP

valves move with them. The valves and buckles are placed in position inside the steam chests, the back ends of the spindles being cottered to the valve rods, which work in circular guides forming part of the motion plate. The accuracy of the work must be such that the centre lines of spindles and guides must either coincide exactly or if, as is frequently the case, there is an offset the centre lines must be perfectly parallel to each other. The spindles must move freely without shake or binding in the guides.

Whilst the above work is being done to the motion, other erectors may be fitting the axleboxes into the hornblocks by filing the side faces until they bear evenly on the horncheeks. The boxes should be a fairly tight fit between the hornblock faces, but not so tight that they cannot move up and down to allow for the vertical movements of the springs. Everything must be perfectly "square," and the axleboxes on one side exactly in line with those on the other, otherwise the axles would be on the "skew," and the locomotive would run badly. The hornblock faces on opposite sides of the engine have previously been finished exactly in line and tested across the engine by straightedges, and the machining and boring of the axleboxes is done so accurately by modern methods that the centre of each axlebox is equidistant from each of the hornblock faces.

If the work has not already been done in the wheel shop, the axleboxes, after having been removed from the hornblocks, are fitted to the

7—(5350A)

88   STEAM LOCOMOTIVE CONSTRUCTION

journals on the axles, to bed down the bearings as described in Chapter VI.

It should be understood that many of the above operations are carried on simultaneously by different gangs of men. For the purposes of quick erection, as many men as possible work together on the same engine under a leading hand or "chargeman," as he is termed. The chargeman is responsible for the whole work on a single engine, and frequently may have two or three engines in hand at once. He is in turn responsible to the shop foreman, who superintends the whole of the work done in the shop, in which a dozen to twenty engines may be in course of erection.

Boiler, Smokebox, etc. The engine is now ready for the boiler to be placed on the frames. The boiler comes in a finished condition from the boiler shop, with all mountings attached, and is lowered on to the frames by an overhead crane. It is at this stage that the accuracy of the boiler shop work is tested finally, for when the firebox has been lowered down into position between the frames, the expansion angle irons K Fig. 2,* must be square so that they rest evenly on the top edges of the frames, whilst at the same time the bottom flange of the smokebox tubeplate or the front end of the boiler barrel, must fit accurately

* The function of expansion angle irons is explained in the primer on *The Steam Railway Locomotive*. Particulars of other details mentioned in this chapter may also be found there.

ERECTING SHOP 89

up to the cylinder flange to which it has to be bolted. The accuracy of the work is generally such that no alteration is needed, but it does occasionally happen that in these respects a boiler is slightly out of truth, and a lot of extra chipping and fitting of the expansion angle iron, and heating and setting of the tubeplate flange are then required.

The smokebox is then built up in position. It is made of thin steel sheets riveted together and to the smokebox tubeplate flanges. The front portion is formed of a large angle iron ring, and a plate, in which the smokebox door-hole is formed, is riveted to this ring. The front plate and the sides at the bottom are bolted to flanges on the cylinders. The smokebox must be quite air-tight, for if it draws air through any crevices or through a badly-fitting door, the boiler will not make steam properly, and any incandescent cinders drawn through the tubes will continue to burn in the smokebox and damage the thin plates. The tops of the (inside) cylinders form the floor of the smokebox, and a covering of fire brickwork and cement is placed over them to protect them from corrosion. The blast pipe and steam pipes in the smokebox are then put into position and jointed up to the cylinders. The steam pipe has also to be jointed with a steam tight joint to the T-pipe in the smokebox. The blast pipe is set accurately with a plumb line, since the centre of the top must coincide with the vertical centre line of the chimney, otherwise a

90   STEAM LOCOMOTIVE CONSTRUCTION

one-sided blast would result, which would soon wear a hole in the chimney ; also, the boiler would not " steam " properly. The fitting of the smokebox door—of which the hinges must be exactly level and strong enough to prevent the door from sagging—together with the fixing of the blower and blower pipe, complete the work in the smokebox.

Meanwhile other men are engaged in fixing the boiler clothing, cab, and splashers. For the clothing, which is essential to reduce the radiation losses from the boiler, the materials in use were formerly boards of yellow pine wood painted with fire resisting asbestos paint. Now felt sheets, asbestos mattresses, or special blocks of magnesia cut to fit the contour of the boiler are generally used. They are secured to a light framework of angle irons and strips fastened round the boiler, and finally covered with thin steel or planished iron sheets screwed on to this framework. The splashers and cab sides are bolted to the platforms, the front plate of the cab resting upon the top of the firebox shell.

Wheels and Springs. The engine is now wheeled. The finished wheels on their axles have already come in from the wheel shop, and are run on to the rails of the pit over which the engine is being built. They remain at the end of the pit until required, the axleboxes having meanwhile been bedded to the journals as previously described. The engine is lifted by the overhead crane, and

ERECTING SHOP 91

the wheels rolled underneath, each pair being placed in position under the corresponding horn-blocks. The axleboxes are placed on the journals and the engine lowered very slowly and carefully, a man being stationed by each wheel to guide the axleboxes into the spaces between the horns.

The springs, if of the type placed above the axleboxes, have already been put into position with the side links and pins in place, so that as the engine comes down the central spring pillars bear upon the axleboxes and the weight of the engine is gradually taken by the springs. The horn-keeps are then bolted into position. If the springs are "underhung," they have to be put into place, after the engine has been lowered on to the wheels, by packing the axleboxes up at a convenient level and then slipping the pins for the links and spring buckles into their respective holes.

Motion Work. To put the connecting rods into place the small end straps and brasses are coupled to the crossheads on each side, and each big end with its brasses is put on to the crank pin journal of the driving axle. As these rods are very heavy, the large end of the rod is made to rest upon a timber baulk placed across the pit, and is then lifted by three men until the butt end slips into the open jaw of the strap, where it is secured by the bolts and cotter.

The eccentric rods and eccentric straps are then coupled up. The eccentric sheaves have already

92   STEAM LOCOMOTIVE CONSTRUCTION

been keyed on at the proper angle to the crank axle in the wheel shop, and nothing remains to be done but to place the straps and rods in position, and couple the latter to the links of the link motion, which are put up at the same time and connected to the valve spindles. The brackets on the frames which carry the reversing shaft are bolted into place temporarily, but are not fixed permanently until the valves have been set, this being the next operation, a description of which is given in the following chapter. For the moment it will be supposed that the valves have been set. The coupling rods are then put on to the pins in the wheels. These rods are formed at the ends with simple circular brass or white metal bushes which are pressed into the holes in the rods by hydraulic pressure. Formerly double brasses secured by cotters, somewhat similar to those used for connecting rods, were used for the coupling rods, but they are now almost obsolete in this country, though still used to a large extent on continental and American engines. To prevent the coupling rods from being thrown off the wheels, washers are placed on the ends of the coupling rod pins, and secured by nuts and taper split-pins.

Only miscellaneous work now requires to be done to complete the engine. The firebars are placed on their bearers in the firebox, and the ashpan is bolted up underneath the firebox. The brake gear has also to be rigged up with all its attendant levers, and the cylinder cocks and

ERECTING SHOP 93

levers, sandgear, sandpipes, various steam and water pipes and connections to the tender are then put up. Finally the buffer plate with buffers, and the drawgear complete the engine which is then ready for trial.

Time Required for Erecting an Engine. The time taken to erect an engine of ordinary size, say with six wheels all coupled, may be taken to be about 45 hrs. from the time the frame plates are placed in position. Large modern engines with ten or twelve wheels will take a correspondingly longer time. The period mentioned would apply to a well organized works in which a large number of standard engines were being erected together, three of which would be under the care of one chargeman. In smaller works about four weeks may be taken for the erection. A few instances of very rapid erection have been recorded in which special preparations were made for the work. In 1888 the London and North-Western erected a standard mineral engine in $25\frac{1}{2}$ hrs., but this record was beaten in 1891 at the Great Eastern Works at Stratford, when a somewhat similar engine was put together ready for trial in the phenomenal time of 9 hrs. 47 min. by forty-four men and boys.* The tender was completed by another gang within the same time. These, however, are quite exceptional cases.

* A full description of this feat was given in *Engineering*, 18th December, 1891.

94 STEAM LOCOMOTIVE CONSTRUCTION

It should be understood that in a well-organized works three or four orders, each for a dozen to twenty engines, are in hand at the same time. Thus the first order will be in course of completion in the erecting shop, whilst the boiler, fitting, and machine shops are dealing with the parts for No. 2 order. The forge and foundry and part of the boiler shop have in hand at the same time the forgings, castings, and boiler plates for the third order, and work may also have been begun in the pattern and template shops on order No. 4.

The Tender. Although the tender is constructed and erected in a separate tender shop, it may conveniently be considered here. The frame plates are of the same quality of steel as those of the engine, their thickness, about $\frac{7}{8}$ in., being slightly less than that of the engine frames. There are usually four frame plates, which are marked off to templates, drilled, and erected with the necessary cross stays in a similar manner to those of the locomotive.

The wheels, axles, axleboxes, hornblocks and springs are also made in a similar manner, the chief difference being that the wheels, being smaller, do not require wheel lathes with such large faceplates. The wheels are simply pressed on to the axles by hydraulic pressure, and are not secured by keys, as are the coupled wheels of the locomotive.

The tank is a separate structure which is placed

ERECTING SHOP

upon and fastened to the framing. The plates, about $\frac{1}{4}$ in. thick for the sides and $\frac{3}{8}$ in. for the bottom, are of mild steel of good quality, but not of the specially high grade which is required for boiler plates. The rivet holes are marked off, drilled or punched, and the plates riveted together with angle irons and supporting stays with $\frac{1}{2}$ in. rivets spaced at about $1\frac{3}{4}$ in. centres.

The tank is bolted to an angle iron, which runs round the top edge of the frame, so that it can be lifted off readily for examination when required.

In erecting the brake gear, both on the engine and on the tender, care must be taken that the brake blocks do not rub against the tyres when the brakes are " off," otherwise considerable heating and undue wear will occur. The blocks are arranged to be just clear of the wheels at the top side, and about $\frac{1}{4}$ in. clear at the bottom. The reason for this is that the brake pull rods are below the blocks, whilst the fulcra to which the brake block hangers are attached are above, and therefore the bottom sides of the blocks move through a greater distance when the brakes are applied.

CHAPTER IX

SETTING THE VALVES

Object of Valve-Setting. The movement of the slide valve with the definitions of "lap," "lead," and "travel" of the valve, together with an explanation of the positions of the piston at which the steam is cut off, the exhaust opens, and compression of the steam begins, are explained in *The Steam Railway Locomotive* and in *Steam Engine Valves and Valve Gears.** The distribution of the steam, and the location of the essential points of cut-off, release, etc., are settled in the drawing office when the engine is designed, but in the erection of the valve gear small inequalities and differences occur. The object of valve setting is to rectify these differences, more especially with a view to producing as nearly as possible an equal distribution of the steam at the front and the back ends of the cylinders. The rectifications are made by slight alterations to the lengths of the valve spindles or eccentric rods.

It should, however, be pointed out that it is impossible to obtain exactly equal distribution of steam at the two ends of the cylinder, owing to the effect of the angular obliquity of the connecting rod. If the connecting rod moved

* Both in the same series as this volume.

96

SETTING THE VALVES 97

bodily in a direction parallel to the stroke of the piston the point of cut-off of the steam would be the same at both ends of the cylinder, but owing to the sloping position which the rod necessarily occupies, the point of cut-off is later on the " out-stroke," when the piston is moving towards the crank axle, than on the " instroke," when it is moving away from the axle. For example, the " valve card " of a certain locomotive shows that when in " full gear " the piston moved $18\frac{3}{4}$ ins. on the out stroke (the total stroke being 24 ins.) before steam was cut-off, but on the return or inward stroke the cut-off occurred when the piston had travelled only $16\frac{1}{4}$ ins.

Systems of Valve-Setting. It is possible to set the valves on three different systems, each being dependent upon one of the three variables which affect the steam distribution. These are—

1. Equalizing the lead at the front and back.
2. Equalizing the port opening at front and back.
3. Equalizing the point of cut-off at the front and back.

The third of these, which involves unequal laps, is rarely adopted in locomotives and need not be further considered. Of the two others the system of equalizing the leads is more usual, as this gives a more nearly equal compression of the steam at the two ends of the stroke.

Bumping Marks and Clearance. If the piston were allowed to travel far enough at each end of

98   STEAM LOCOMOTIVE CONSTRUCTION

the cylinder, it would "bump" the cylinder covers. The engine is so designed that this does not take place, and a space of about $\frac{1}{4}$ to $\frac{3}{8}$ in., known as the "clearance," is allowed between the piston and the cylinder cover at each end of the stroke. Since the movement of the piston and of the crosshead connected to the piston rod are exactly the same, all measurements of piston stroke may be taken from the crosshead relatively to the slide bars. The "bumping marks" are scribed on the slide bars at each end before the connecting rod is put up, by pushing the piston by hand to each end as far as it will go until it strikes the cylinder covers, and marking the positions of one end of the crosshead slide block on the bars. When the connecting rods and motion have been put up, and the wheels revolved as described below, the terminal positions of the slide blocks should be distant from the bumping marks at each end by the amount of the clearance.

To set the valves it is found more convenient to turn the wheels by means of the following simple apparatus than to move the engine backwards and forwards on a length of line outside the shop. Two brackets are fixed beside the rails, one on each side of the pit over which the engine is built. These brackets have bearings in which rollers are fitted, and the engine is lifted slightly so that the circumferences of the driving wheels rest upon the rollers. The rollers are revolved by means of a shaft which lies across the pit, the shaft being provided with a ratchet

SETTING THE VALVES 99

or four long lever arms, the latter being pulled round by a man stationed in the pit during the valve setting process. The rollers cause the driving wheels to revolve, only a small portion of a revolution being necessary at one movement during valve-setting. The wheels must be turned in the proper directions when setting the valves for forward or back gear.

Setting the Valves. Before the eccentric rods are coupled up to the expansion links, the points at which the valve closes the ports to steam are marked upon the valve spindles. A very thin piece of tin A_1 (Fig. 43–I) is held in the front port, and the valve with valve rod is moved carefully against it. The valve spindle S is then marked at a_1 outside the steam chest by means of the trammel T, the straight pointed end of which is placed in a centre " pop " made in the back of the steam chest cover, whilst the right-angled end scribes the mark a_1 on the valve spindle. A small " pop " is made with a sharp punch to show the position of this mark. Similarly the point a_2 at which the valve closes the port at A_2 at the other end (Fig. 43–II), is marked, by transferring the tin to the latter port and pushing the valve in the opposite direction against it. The two " pops " a_1 and a_2 on the spindle show the positions of the latter when the steam is cut off, since these are the points at which the valve closes the ports.

The eccentric rods are then coupled up, and the " dead centres," i.e., the exact points at

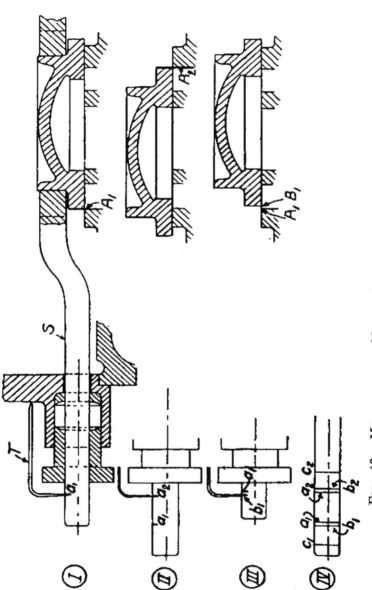

FIG. 43.—MARKING THE VALVE SPINDLE FOR VALVE SETTING.

SETTING THE VALVES 101

which the piston reaches the two ends of its stroke, have to be found. It may be noted here that there are four dead centres, two for each side of the engine. As the valve-setting for one side is the same as for the other, it will be sufficient to describe one side only.

The driving wheel (Fig. 44) is turned in the direction of the arrow for forward motion until the slide block D is nearly, but not quite, at the front end of its stroke as shown in the illustration. The crank will then occupy the position shown at C_1. At the extreme end of the stroke the edge of the slide-block would reach the dotted line A. A centre punch mark P is made on the side of the slide block, and another similar mark S on the side of the slide bar. A pair of compasses is then set to the distance PS. A centre punch mark is made on the engine frame at Q near the driving wheel, and a mark m_2 is made by a pair of compasses or a trammel on the tyre. The wheel is then slowly revolved so that the slide block D reaches the end of its stroke A, and then begins to move in the opposite direction towards the crank. A man in the pit keeps the compasses set to PS with one point in S and as soon as the other point enters the pop mark P the rotation is stopped. With the trammel having one point in Q, a second mark m_2 is made on the tyre, the first mark having moved on to the position m_1, as the crank has moved round from C_1 to C_2. The distance $m_1 m_2$ on the tyre is divided equally, and a " pop " is made in the mark M, the centre

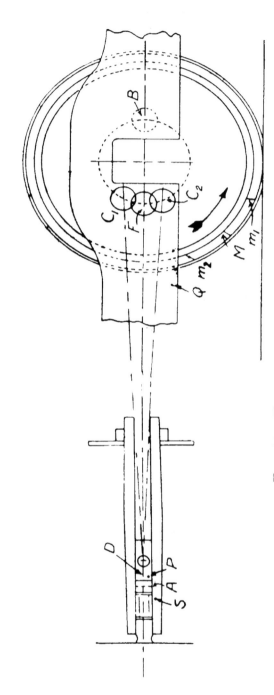

Fig. 44.—Finding the Dead Centres.

SETTING THE VALVES 103

line between m_1 and m_2. The wheel is then moved back to the position C_1 of the crank, and again moved forward until the trammel with one leg in Q exactly reaches the pop M with its other pointed end. The point M will now occupy the position which m_2 has in the illustration, and the position of the crank will now be at F', the edge of the slide block being at the end of the stroke A. This is the front dead centre. The back dead centre B is found similarly by turning the wheel through half a revolution, and two further dead centres are also marked for the crank on the opposite side of the engine.

It may be added that the above somewhat elaborate method is necessary, because it is practically impossible to find the dead centre sufficiently accurately by merely watching the crosshead and marking the point when the latter begins to reverse its direction of motion. The motion of the crosshead near dead centre is very slow, and the crank will move through a small angle without apparently moving the crosshead.

The engine having been set on front dead centre with the crank at F (Fig. 44), by setting the wheel trammel in the pops Q and M, the front port should be open a slight amount $A_1 B_1$ (Fig. 43–III). This amount is the lead at the front end, and is marked with the valve trammel on the valve spindle at b_1. The distance $a_1 b_1$ is therefore equal to $A_1 B_1$. These marks on the spindle are also shown in Fig. 43–IV, and a similar mark b_2 is also

8—(5350A)

104 STEAM LOCOMOTIVE CONSTRUCTION

made for the lead at the opposite end, when the crank is on back dead centre in the position B (Fig. 44). The lead at the back end will therefore be equal to the distance between the marks $a_2 b_2$ on the spindle (Fig. 43–IV). In some works the leads are measured, not by marking the valve spindle, but by pressing thin wedges into the space $A_1 B_1$ between the edge of the valve and that of the port, so that these edges leave marks on the wedges which can be measured.

If the front lead $a_1 b_1$ is equal to that $a_2 b_2$ at the back the valves are properly set " to equal leads," and no further adjustment is necessary. If, however, the lead $a_1 b_1$ in fore gear at the front is say $\frac{1}{16}$ in, whilst $a_2 b_2$ at the back is say $\frac{3}{16}$ in., the valve requires adjusting. To do this the fore gear eccentric rod is taken down and " jumped " or shortened in the smiths' shop by $\frac{1}{16}$ in. This increases the opening $A_1 B_1$ or the distance $a_1 b_1$ at the front end from $\frac{1}{16}$ to $\frac{1}{8}$ in., and decreases $a_2 b_2$ from $\frac{3}{16}$ in. to $\frac{1}{8}$ in. so that the leads are made equal. Similarly if the lead at the back end be smaller than that at the front, the eccentric rod has to be lengthened slightly. Some engines are arranged with valve spindles having a screw and nut, which can be adjusted to lengthen or shorten each spindle instead of altering the length of the eccentric rod.

The maximum port openings, when the valve is at the ends of its stroke, are also noted and marked on the valve spindle. The valve spindle is blackened by soot from a lamp, so that the marks

SETTING THE VALVES 105

made by the trammel may be seen easily. As
the ends of the valve travel are approached,
when the wheels are rotated, the trammel T is
placed as shown in the pop in the steam chest
cover (Fig. 43–I) and the vertical point is passed
and repassed across the valve spindle as the
latter moves until the latter reverses its direction
of movement. The limiting position gives the
line c_1 (Fig. 43–IV) marked on the spindle, and the
distance $a_1 c_1$ is the maximum port opening at the
front end. Similarly $a_2 c_2$ is the port opening
at the back end. When valves are set to equal
port openings instead of equal leads, the adjustments are such that $a_1 c_1$ is made equal to $a_2 c_2$.
In this case there is a slight difference in the leads
at the front and back ends.

It may be noted that as the reversing lever is
" notched up " when running, in order to expand
the steam in the cylinders, the lead increases
with a Stephenson link motion having " open "
eccentric rods.* For instance an engine, which
has the valves set to equal leads of $\frac{1}{8}$ in. when in
full fore gear, may have a lead of $\frac{1}{4}$ in. when fully
notched up, and moreover the link motion may
be so suspended that the leads at both ends are
not equal when the reversing lever is in this latter
position. It is therefore usual for the valves
to be set to equal leads, when the reversing gear
is placed in the notch or position in which the

* For further particulars of Stephenson's link motion
with open and crossed rods, and the effect of the latter
on the lead, see *Steam Engine Valves and Valve Gears*
in this series of primers.

106 STEAM LOCOMOTIVE CONSTRUCTION

engine runs most of the time, instead of setting them with the lever in full fore-gear.

The valves are also tried for " back-gear," in which the engine runs tender first. In the case, however, of tender engines the running done tender first is comparatively small in amount and the setting of the valves in back-gear is sometimes left, unless the inequalities between the front and back of the cylinders is very pronounced, in which case some adjustment is made to the back eccentric rod which may involve a slight alteration of the steam distribution in fore-gear.

Finally, when the valves have been set, the positions of the crosshead, which are equivalent to those of the piston, are measured on the slide bars corresponding to the four points : (1) steam cut-off ; (2) exhaust opens ; (3) exhaust closes ; and (4) compression begins. Owing to the obliquity of the connecting rod these positions are different for the outward and inward strokes of the piston. A record of these measurements is kept for every engine.

CHAPTER X

INSPECTION AND TESTING

THE testing work in connection with locomotives may be divided into two main categories—

(*a*) The testing of the materials of which the engine is constructed.

(*b*) Testing and trials of finished locomotives.

Testing and Inspection of Materials. This is done at the works of the various manufacturers of the plates, bars, axles, tyres, etc., who provide facilities in the shape of testing machinery, in which specimens cut from the material are broken and their physical behaviour noted. The railway companies, if the materials are for use in this country, or the consulting engineers, if the engines are being built for abroad, have their representatives or inspectors at the works of the manufacturers, to pass the materials for acceptance if the tests prove satisfactory, or reject them if they do not come up to the specified requirements.

Tensile Tests. The most important physical test is known as the tensile test. A piece of the plate is machined to definite rectangular or circular form and dimensions, and placed in the jaws of the testing machine. The specimen is

108   STEAM LOCOMOTIVE CONSTRUCTION

then pulled by the direct action of a hydraulic cylinder, the pull being counterbalanced and measured by a long steelyard on which there is a movable weight. After a certain time the piece begins to stretch, and the first sign of this is measured on the piece by a pair of compasses which have been set to a definite and specified length varying from 2 to 10 ins. according to the test specimen. This length is marked on the specimen before the test is begun. The point at which the first signs of permanent stretch are observed is termed the *yield point*, and the stretching denotes that the limit of perfect elasticity of the material has been passed. The position of the weight on the steelyard at this moment is recorded, after which the pulling is continued until the specimen breaks, the position of the weight being then noted again. The steelyard is marked directly in tons, so that the position of the weight shows without further calculation the total number of tons pull on the specimen. The number of tons divided by the known original area of the section of the specimen in square inches, gives the yield point or the ultimate breaking strength of the material in tons per square inch.

Just before the specimen breaks it pulls out somewhat after the manner of a piece of toffee. The extension of length over that of the specimen, as it was originally, and also the contraction of area where it has broken, are also recorded, and give a measure of the ductility of the material.

INSPECTION AND TESTING 109

Table I shows the ultimate tensile strength and minimum permissible elongation for the most important materials used in locomotive construction.

TABLE I—STRENGTHS OF MATERIALS FOR
LOCOMOTIVE CONSTRUCTION

Name and use of Material.	Ultimate tensile strength tons per sq. in.	Percentage elongation above original length not less than
Mild steel boiler plates . .	26 to 32	22% in 8 ins.
Mild steel boiler rivets . .	24 to 28	27% in 8 diameters
Mild steel frame plates . .	28 to 32	20% in 8 ins.
Steel crank axles . . .	not less than 30	20% in 2 ins.
Steel crank axles, oil treated.	not less than 35	20% in 2 ins.
Steel straight axles . . .	25 to 40	25% to 20% in 2 in.
Steel tyres, class C . . .	50 to 55	13% to 11% in 2 ins.
Steel tyres, class D . . .	56 to 62	10% to 8% in 2 ins.
Steel castings with wearing surfaces	not less than 35	10% in 2 ins.
Cast steel wheel centres. .	not less than 26	15% in 2 ins.
Copper plates for fireboxes .	not less than 14	35% in 8 ins.
Copper stays for fireboxes .	not less than 14	40% in 8 diameters
Mild steel tubes for boilers .	not more than 24	28% in 8 ins.
" Best Yorkshire " iron bars, 1 in. to 4 ins. diameter .	21 to 23½	varies according to bars.

Spring steel is not usually tested for tensile strength, but the springs are subjected to rapid deflection tests.

Bending Tests. Other tests specified include " cold bend " and " temper bend " tests. In the former, specimens cut from plates are doubled over cold, until the internal radius is not greater than the thickness of the plate, and the test piece must withstand this without fracture. The " temper bend " test is similar, but the piece is heated beforehand to a red heat and quenched

110   STEAM LOCOMOTIVE CONSTRUCTION

in water at a temperature not exceeding 80°F. Rivets are tested by being bent cold until they are completely doubled up without fracture, and the heads must be hammered flat without cracking.

The specified tensile and bending tests differ according to the material concerned. In the case of plates there is always sufficient excess material at the sides from which a test specimen can be cut, so that each plate can be tested if required. In the case of straight axles and tyres which are tested to destruction the inspector selects any two per cent. of the number ordered, and the quality of the whole lot is decided by the behaviour of those so selected.

Impact Tests. In addition to tensile and bend tests, axles and tyres are also subjected to the " falling weight test." The test axle is placed upon bearings which are placed at a distance apart varying from 3 ft. for an axle 4 ins. diameter at the centre, to 5 ft. for axles 6 ins. or more in diameter, and a weight of 1 ton is dropped on to the centre of the axle from a height varying from 16 ft. for the 4 ins. axle to 35 ft. for the 6 ins. and larger axles. The axle must withstand five such blows without breaking. Under the first blow it bends, and is then turned over to receive the next blow, this being repeated each time. If it stands this test the axle is finally broken.

Tyres are tested in a similar manner by being placed on edge, a 1-ton weight being then dropped

INSPECTION AND TESTING

on to them from heights of 10, 15, and 20 ft. until the tyre deflects under the blows a specified amount without breaking.

Chemical Tests. Each railway company has, in addition to a mechanical testing shop, a well equipped laboratory in which samples of the above materials are analyzed. Chemical tests are not often specified for the materials, beyond limiting the amounts of sulphur and phosphorus in steel, both of which are deleterious. Thus boiler plates must not show more than 0·05 per cent. of either of these elements, but in the case of springs, tyres, and axles, these limits are reduced to 0·035 per cent. Copper firebox plates are specified to contain from 0·25 to 0·55 per cent. of arsenic, which has the valuable property of reducing wear in service.

Trial Trips of Completed Locomotives. Contrary to what might be supposed, new locomotives do not, except under special circumstances, undergo any elaborate tests, such as are usual with large marine or mill engines. An engine built at a railway company's works is immediately sent out on a trial run of some 20 to 25 miles without a train. During the trial the working of the motion, springs, brakes, injectors and sandgear is noted carefully. The big end and axlebox bearings must run perfectly cool, and there must be no leakages at any of the joints, to which the fittings and mountings on the

112   STEAM LOCOMOTIVE CONSTRUCTION

boiler are attached. Everything that is not satisfactory is noted for rectification when the engine returns. A second trial may occasionally be necessary, especially in the case of an engine of a new class, and finally, when everything is satisfactory, the engine is sent to the paint shop to be painted.

An express engine is not at once set to work on fast trains, but for about three weeks it is employed on local stopping passenger trains at the shed where the works are situated. This gives the engine an opportunity of " finding its bearings," and any defects which may show themselves are rectified. Afterwards the engine is tried for a week or two on fast trains, and is then ready for regular service and is sent away to its allotted station. An engine which has come out of the works after repairs undergoes exactly the same routine.

Special Tests in Service. These are usually " road tests," that is, they are made in service on the line when the engine is working its usual trains. Generally such tests consist of a long daily series, extending over a period of several weeks, and they are generally undertaken, when it is desired to estimate the coal and oil consumption of a new or altered class of engine in order to compare the results with those given by an engine of an existing class, which has hitherto been working the same trains.

For this purpose records are kept of the weight

INSPECTION AND TESTING 113

of each train ; of the miles run, and the time gained or lost, from which the actual average speeds can be calculated ; and of the amounts of coal burnt and oil used. The data obtained are tabulated each day, and the results are compared carefully. Minor alterations in the new engine may be made as a result of these trials. British engineers generally prefer this form of extended road test, since it gives results under commercial working conditions. The tests may be elaborated by including a dynamometer car in the train, attached immediately behind the tender, so as to secure a continuous record of the pull of the engine on the tender draw bar. At the same time indicator diagrams are taken at intervals. The indicated horse power calculated from the indicator diagrams, when compared with the useful horse-power calculated from the draw-bar pull registered by the dynamometer car, gives a measure of the efficiency of the engine.

Tractive Force and Horsepower. It may be explained that the power of a locomotive is not, as in marine and stationary engines, estimated on a horsepower basis, but by the tractive force.* The horse-power unit involves the speed of the train, which varies continually. The resistance due to the load also varies continually with the gradients and curves of the line, and with the wind. The horse-power at any given moment can, of

* An explanation of the term " tractive force," is given in the primer *The Steam Railway Locomotive.*

FIG. 45.—LOCOMOTIVE TESTING PLANT.
G.W.R. Works, Swindon.

INSPECTION AND TESTING 115

course, be calculated from the indicator diagrams, but it varies so greatly throughout the journey that this unit is not a convenient one for locomotives.

Coal Consumption. The coal consumption varies greatly with the class of engine, weight of train, gradients, weather, and speed. An express engine may burn anything from about 23 to 50 lbs. of coal per mile. Experiments on the Midland Railway, the main line of which has heavy gradients, showed an average consumption during certain tests of from 0·07 to 0·16 lbs. per ton-mile with different engines and trains.

Testing Plant. Another and modern form of test may be mentioned briefly. This is made in the works on a test plant, the conditions being such as to approximate as nearly as possible to a road test. The engine is placed over a special pit in such a way that its coupled wheels, drive by friction another set of wheels and axles fixed across the pit. The engine can be run under its own steam at any speed desired, and the draw-bar pull is measured by a special apparatus at the back. · Coal, water and oil consumptions at certain speeds over a definite length of time are measured and recorded. This form of test plant, which is extremely valuable for experimental purposes is of American origin, but there is only one in this country, at the Great Western works at Swindon. An illustration of this is given in Fig. 45.

CHAPTER XI

LOCOMOTIVE MAINTENANCE AND REPAIRS

THE locomotive, when in traffic, is under the charge of one of the running sheds or districts, into which the railway is divided for locomotive running purposes. The usual practice is for ordinary running defects and minor repairs to be attended to in the small workshops attached to the sheds. The facilities provided for this purpose depend upon the size of the running shed, and in some important sheds many "heavy repairs" are now executed. Generally speaking, running shed repairs include re-turning the tyres, refitting axleboxes and bearings, repairs to the motion and brake gear, and the fitting of new springs. The springs and other parts are, of course, supplied from headquarters. Minor boiler repairs such as the replacement of stays are also done in the running sheds, but when the boiler requires heavy repairs the engine is sent away to the principal works.

Records are kept of the mileage and of every repair done to the engine, however small. Further, periodical examinations are made of most of the important parts of the engine after it has run a definite number of miles, or for a certain length of time. Thus the firebox would be examined about once a month, and the tyres and axles of express engines after running about 4,500 miles,

116

LOCOMOTIVE MAINTENANCE AND REPAIRS 117

but bogies, springs and slide valves would be allowed to run 20,000 miles between each periodical examination. Before the engine is sent into the principal shops for heavy repairs, a special report is sent there some weeks in advance, giving full particulars of the condition of the principal portions, so that the new parts are ready and no unnecessary delay occurs after the engine has been sent in.

An engine may be in service from one to two years between its visits to the principal shops. The time depends upon the class of engine, the service upon which it has been employed, the mileage run, and the general condition in which it is.

Boiler. The boiler requires more supervision than any other part of the locomotive, and is the most expensive in the matter of upkeep. Sir John Aspinall, an eminent locomotive engineer, once made a statement* which will be endorsed by every locomotive man that " there was only one thing which caused trouble, namely, the boiler. The engine part was quite satisfactory, and never gave any trouble, but the boiler was an everlasting trouble."

To make a thorough examination of the boiler, the tubes are withdrawn, and the incrustation due to the water is removed from the plates by means of suitable chipping tools, in the use of

---

* Proceedings Institution of Mechanical Engineers, July, 1909.

118   STEAM LOCOMOTIVE CONSTRUCTION

which great care must be taken not to " nick " or damage the plates, or subsequent fractures are liable to result.

The steel boiler and firebox casing plates are liable to corrosion and " pitting," cracks, and " grooving." The corrosion takes the form of a uniform wasting of the plates, whereby they become thinner, and eventually unable to withstand the pressure required. This form of corrosion is usually found on a longitudinal belt from 1 ft. to 2 ft. wide near the water level. Pitting is the more usual form of corrosion, in which the plate is honeycombed with small cavities either isolated, or running into one another to form depressions of considerable size. This defect is probably caused by combined chemical and galvanic action due to dissolved acids in the water. The scale deposited by the feed water, if thin, helps to protect the plates, but the alternate expansion and contraction of the plates, as they are heated and cooled, helps to detach pieces of the scale, leaving the plates exposed to the action of any acid in the water. It may here be remarked that most boiler and firebox troubles are caused by the alternate expansion and contraction of the different parts, and that its wear and tear are due to causes quite different from those which operate in the cylinders and motion. " Grooving," usually found on the smokebox tubeplate and at the foundation ring which unites the copper firebox to the firebox shell, is also due to the same cause. As the tubes expand they tend to push out the centre of the tubeplate, whilst it is rigidly held at the edges, especially where it is jointed to the flange of the cylinders. The constant bending and unbending causes a slight crack in the plate which ultimately develops

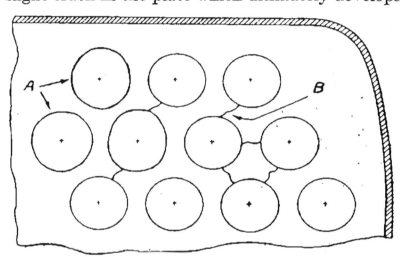

Fig. 46.—Oval Tube Holes (A) and Cracked Tube Plate (B).

into a groove, the latter being enlarged by the action of acid in the water.

A badly grooved or wasted plate is renewed, but if the defects are not so serious, the plate is repaired by riveting a patch over the defective portion. Rivet heads are liable to corrosion, and defective rivets are knocked out and replaced.

Firebox. The firebox is particularly subject to defects. The copper plates, especially at the firehole or at the flanges where the plates are united, are wasted away by the action of the flame.

120   STEAM LOCOMOTIVE CONSTRUCTION

Cracks and fractures as at *B*, Fig. 46, are frequently found across the tubeplate between the holes in which the tubes are secured ; such cracks also appear between the stay holes in all the plates. As these cracks extend rapidly great care is taken in looking for them. As the tubeplate expands in a vertical direction when the boiler warms up, the tube holes gradually assume an oval shape and leakage then occurs round the tubes. Overheating of the plates, due to occasional shortness of water, causes burning of the plates. Expansion and contraction cause frequent breakages of the stays which unite the roof and sides to the firebox casing, and bulging-in of the plates may then result. The presence of defective side stays is detected by tapping the heads with a light hammer, when an experienced man can tell them by the sound. Stay heads are generally burnt away by the flame.

All these defects, of which some are sure to be found, would become dangerous if left, and for this reason the firebox is thoroughly examined at frequent regular periods. The high pressures used in modern locomotives, from 170 to 225 lbs. per sq. in., increase the tendency to firebox troubles, and some engineers take advantage of the benefit due to super-heating by reducing the pressure to 160 lbs. per sq. in.

To repair cracked plates, patches of various sizes and shapes are used. It is essential in patching to cut away the defective portion of the plate first. This leaves an oval or rectangular hole, and a piece of new copper plate is marked off and cut out slightly larger than the hole, so that it forms a lap all round. The patch is then carefully bedded down and the stay holes and rivet holes marked off in the new piece after

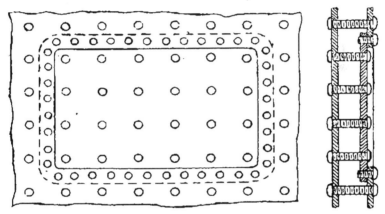

FIG. 47.—PATCH ON FIREBOX PLATE.

which it is riveted or fastened by stud bolts according to the position which it occupies, for in some parts of the firebox riveting cannot be done. It should be noted that such patches are never put over the cracks in the old plate, but the latter is always cut out, for the reason that were the old plate left there would, after the patch has been put on, be a double thickness of plate, which would reduce the conductivity, so that overheating and burning would result.

If a firebox plate be in bad condition, the lower half may be cut away, and a new half-side, or half tube plate may be riveted to the old upper half.

When tube holes are oval as at *A*, Fig. 46, and the plate between the holes is cracked as at *B*, the holes are enlarged with a "rose-bit" tool, and specially turned and screwed plugs are inserted tightly into the holes. These plugs are then riveted over on each side of the tube plate to cover the cracks in the plate as far as possible. The plugs, if not left solid, are then drilled to receive new tubes, which are somewhat smaller than the tubes which were previously in these holes. Fig. 48 shows two of these bushed holes in section.

Fig. 48.—Bushed Tube Holes in Tube Plate.

Every five or six years an entirely new firebox is required. This involves stripping the whole of the interior of the boiler, cutting through and knocking out all the stays, and removing the foundation ring. The boiler, when undergoing this repair, is turned upside down and the new firebox put in as in a new boiler (see Chapter II). The whole boiler will require renewal after 12 to 16 years; this is a comparatively simple matter, the old boiler being lifted away from the frames, and the new one placed on them as described in Chapter VIII.

Firebox Stays. Leaky and broken firebox stays are a constant source of trouble. Leaky copper stays, if not in too bad a condition, may be attended to in the running shed, the heads being lightly riveted over and "caulked" with a light tool which has a circular head. The principal trouble, however, is with broken stays. This defect is shown in Fig. 49. The copper or inside firebox expands in a vertical direction as the steam is being raised, and rises so that the stay is bent or inclined as shown at *A*, since the outer steel firebox shell to which the stays connect it does not rise so rapidly or to the same extent. The constant bending and re-straightening of the stays causes them to crack as at *B* and finally fracture as at *C*. When putting in new stays the holes are slightly enlarged, tapped for new screw threads, and fitted with correspondingly larger stays.

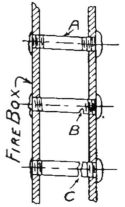

FIG. 49.—CRACKED AND BROKEN STAYS, WITH HEADS WASTED AWAY INSIDE FIREBOX.

Tubes. The chief defects of steel tubes are (1) pitting of the outsides due to the same cause as pitting of the boiler plates, and more particularly (2) leakage at the firebox tube-plate. In former years leakages were stopped temporarily by driving a conical tool, known as a "drift," into the open firebox-end of the tube. This forced the metal of the tube against the walls of the tube-holes, but was very injurious to the tube-plate, and its use is now forbidden. Instead,

124 STEAM LOCOMOTIVE CONSTRUCTION

a special " tube-expander " is used. This consists of a casing which will enter the tube, and in it are slots through which a number of small rollers protrude slightly, the axes of the rollers being nearly parallel to the axis of the tube. An internal conical mandril can be pushed in by the operator so that the cone gradually forces the rollers outwards against the inside wall of the tube, and as the tool revolves the rollers roll the metal of the tube against that of the tube-plate, thus making a tight joint. In some cases a thin soft copper ring or ferrule is placed between the outside of the tube and the hole in the tube-plate, the ferrule being practically squeezed between the tube and the plate. The outer ends of the tubes project into the firebox about ⅜ in. and are afterwards " beaded " or worked over against the face of the tube-plate.

Safety Valves. These and other steam valves wear slightly on the valves and seats, so that they leak and have to be " ground in." A little fine emery and oil is sprinkled on the valve and its seat, and the valve is rotated backwards and forwards until both are ground together to a perfect seating, which can be seen from the dark colour all round their bearing faces.

Frames. Frames occasionally give way by breaking from the inside corners of the recesses into which the driving hornblocks fit, as at *A* in Fig. 17. The alternate thrust and pull due to the steam action is felt most severely at these

LOCOMOTIVE MAINTENANCE AND REPAIRS 125

corners. To repair them the frames may be either patched or welded. A patch is a piece of plate about $\frac{7}{8}$ in. thick cut out to embrace the upper portion of the hornblock and cover the fracture. Welding is now becoming more usual either by the oxy-acetylene or the electric arc process. In both processes, the crack is chipped out, and a V groove formed, into which molten metal is deposited by the flame or arc to produce a solid "weld." Great care has to be taken, especially with the oxy-acetylene process, that the contraction of the frame when cooling after the operation does not cause a new breakage.

Cylinders. The defects which arise in cylinders are principally oval wear of the bore, wear of the port faces, over which the valves work, and cracked or broken cylinders. In the case of cylinders with circular ports for piston valves, such as those shown in Fig. 22, these ports are provided with internal circular liners, through which the necessary port holes are cut. As these wear through the action of the rings of the piston valves, they can be replaced by new liners, and the cylinder casting itself remains uninjured and intact.

For refacing the port faces of slide valve engines, and for reboring the cylinders themselves, the latter are not taken out of the engine frames. To do this would be too expensive an operation. Portable facing and boring machines are used which are fixed in front of the cylinders, the buffer beam having been taken down. These

126   STEAM LOCOMOTIVE CONSTRUCTION

machines were formerly worked by hand, but now an electric motor actuates them through suitable gearing. The bores are carefully callipered, and just sufficient metal is removed to make them truly circular. New piston heads of greater diameter are required when the cylinders have been re-bored.

For cracks and breakages the cylinders must be removed from the frames, and if the cracks are serious the cylinders are replaced by new ones. Some cracks on the exterior are patched with gun metal patches cast from a wooden pattern specially made to fit the contour of the defective part. If the tops of the cylinders are corroded by the smokebox ashes, which always contain sulphur, a patch of copper plate is made to fit and bolted on. If the port faces are badly worn they are faced-up and plates or false faces are screwed on, which are machined to the original dimensions. Cracked " bridges " between the ports may be repaired by cutting away the metal on each side of the crack and dovetailing a piece into the space formed.

Acetylene welding has been used, chiefly in America, for repairing cylinders, especially in cases where flanges are broken.

Wheels and Axles. With the exception of the boiler these require more attention than any other part of the engine. The wheel " centres " themselves very rarely crack now that they are of cast steel. The former wrought iron driving wheels used sometimes to crack between the spokes at the boss, but these are not made now, though a considerable number are still in service. Broken or flawed tyres are replaced immediately, no attempt being made to repair them by welding. The chief defect in tyres is the ordinary wear in service, through which the "tread" on the

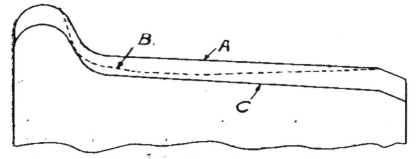

FIG. 50.—TREAD OF TYRE SHOWING WEAR AND SECTION AFTER RE-TURNING.

rail decreases in diameter. The flange therefore becomes deeper relatively to the tread as shown by the dotted line B, Fig. 50, and when it becomes too deep it is liable to catch in crossings on the line and so become a source of danger. The tyres are not removed from the wheels, but the whole set is put into a wheel lathe and re-turned. The re-turning has also to be done when flat places occur on the tyres. Very powerful wheel lathes are required, since a number of exceedingly hard spots are always found on the treads of tyres that have been some time in service. An unavoidable but considerable waste of metal occurs when tyres are re-turned. In the first place

128   STEAM LOCOMOTIVE CONSTRUCTION

much metal has to be turned off to produce the
new proper contour C of the tread and shape of
flange, and secondly, in the case of engines with
four or six wheels coupled, the wheels wear down
to unequal diameters, so that when the smallest
wheel of the set of say six wheels has been turned
down, all the other five wheels must be turned
to exactly the same diameter, and more metal
has to be removed than would be necessary if
each wheel were independent. Coupled wheels
of even slightly unequal diameters cause a great
strain on the coupling rods, and also produce a
" nosing " or side to side action of the engine
when running. In Fig. 50 A is the contour of the
tread of the tyre when new. This wears down
to line B, and the tyre is then re-turned to the
new contour C.

Tyres are usually about 3 ins. thick when new
but, owing to wear and re-turning, they eventually
attain a thickness which will not allow them to be
safely re-turned. Wheels up to about 5 ft. 6 ins.
may be re-turned. so long as the tyres are in
sound condition and provided that the tyres will
not be less than $1\frac{1}{2}$ in. thick after re-turning
but, for wheels larger than 5 ft. 6 ins., the limit of
thickness after re-turning is about $1\frac{3}{4}$ ins.

Axles, especially driving crank axles, are
subject to very heavy bending and twisting forces,
and are watched very carefully for flaws and
cracks. If a minute crack begins to show itself
the axle may be allowed to run until the crack
develops, when the axle is immediately

condemned, the wheels forced off by hydraulic pressure, and a new axle substituted. Other than flaws, the chief running defect is oval wear of the journals which work inside the axleboxes, and a similar oval wear of the crank pins of driving axles. To restore the journals to a true circular form the wheels are put into a wheel lathe, and a very light cut is taken off the axle journals. For truing up crank pins of crank axles, a portable machine is clamped to the crank webs. The axle remains stationary, and the cutting tool revolves round the crank pin.

Axleboxes. These wear on the flanged sides which are constantly moving up and down in the hornblock guides under the action of the springs. The thrust and pull of the rods on driving and coupled axles accentuate this wear on the axleboxes of such axles. As the hornblock guides also wear, the latter are faced up first. This may be done in place on the engine either by filing them up, using a surface plate to detect the high places, or by means of a portable facing machine, or the horn blocks may be removed and re-ground. The axleboxes are then "lined up," i.e. the deficient width across the faces of the hornblocks is made up by riveting a brass liner on one side or by white metalling this side to the proper thickness. The side is then planed and the box fitted into the hornblocks. The crown of the brass which bears on the journals will be worn down and the box will generally require a new brass ; if not too

130   STEAM LOCOMOTIVE CONSTRUCTION

badly worn, the old brass may be re-metalled with white metal. The centres are marked off when the boxes are in place on the engine, and the boxes are then sent to the machine for re-boring. The centres must be properly located so that the distance apart of the different axles is exactly correct.

The Motion, etc. Connecting and coupling rods are examined for flaws, and if any are found in the bodies the rods are at once condemned. Usually flaws in connecting rods occur in the " big-end " straps at the corners and bolt holes. These straps being smaller independent pieces are not so expensive to renew. The rods are also tested for straightness with a straight-edge. The straps will require filing up and perhaps closing a little at the jaws, and the brasses, frequently new, must be a perfect fit in the straps, and be made perfectly square with the rods. The brasses having been fitted and bolted up the ends are placed on the rods and the length between centres of the big and small ends on each rod is trammelled to the dimensions shown on the drawings. The centres are marked and the ends sent to the machine to be bored out to the size of the crank pin journal, to which they are then fitted. Coupling rods are more easily repaired by having new white metal circular bushes pressed into the solid ends of the rods. This work is done in a small hydraulic press.

Slide bars are re-ground on a grinding machine,

LOCOMOTIVE MAINTENANCE AND REPAIRS 131

and set up on the engine in a similar manner to that for a new engine in the erecting shop. Slide valves are re-faced, or if they are worn too thin or are otherwise defective they are renewed. Piston rods and valve spindles require re-turning and re-grinding, as they wear unevenly in the glands, and are often found to be badly scored and grooved. Screw threads and cotter holes in them are examined for flaws, which if found, cause the rods to be condemned.

The various pins in the valve gear wear slack in their holes, and the holes in the links through which they pass are found to be enlarged. The metal being case-hardened these holes cannot be re-bored, but they may be made true by lapping them out with a lead mandrel using emery and oil. An alternative and more rapid method now generally employed is to use a special grinding machine with a very small wheel. If the enlarged holes be used as they are, new pins are required, but in many cases the holes are bushed with hardened steel bushes. The wearing surfaces of the quadrant links are re-ground or lapped out. Great care should be taken that all oil holes are in their proper positions and thoroughly cleaned out.

The brake-gear is overhauled and adjusted and new cast iron brake blocks are provided. There are various other details too numerous for description in this primer, and it remains only to be mentioned that the putting together again of the engine is very similar to the erection of a new engine as described in Chapter VIII.

INDEX

ASSEMBLING the boiler, 18
Axle-boxes, 67, 69, 87, 88
——, defects and repairs, 129, 130
Axle lathes, 57, 59
Axles, defects and repairs, 128, 129
——, built-up crank, 58, 60
——, forging, 36–40
——, machining crank, 58–60
——, —— straight, 56–58

BLACKSMITH'S shop, 42
Blast pipe, 89
Boiler barrel, 8
——, clothing, 90
——, defects in service, 117–119
——, erection on frames of, 88
—— mountings, 24
—— plates, 8, 9
—— riveting, 19
—— shell-drilling machine, 14, 15
—— testing, 24
—— tube and back plates, 15
Brake gear, 95
Brass foundry, 34
Buffer forgings, 40
Bumping marks, and clearance, 97, 98

CASEHARDENING, 78, 79
Castings, 27–29
Coal consumption, 115
Connecting and coupling rods, 76–78, 91, 92
—— —— ——, defects and repairs, 130, 131
Crank pin "quartering" machine, 64
—— pins, 64
Crossheads, 74, 75

Cylinder boring and drilling machine, 53, 54
—— castings, 31–34
—— double-boring machine, 51, 52
Cylinders, 31–34, 49–55
——, defects and repairs, 125, 126
——, erection of, 82–84
——, marking off, 49, 50
——, planing and boring, 51–55

DEAD-CENTRES finding, 101–103
Dome and safety valve seatings, 20
Drilling machine for boiler plates, 11, 12

FIREBOX, 119–120
—— patching, 120–121
—— stays, 22–23
—— ——, defective and broken, 122–123
Forging and smithing, 35
—— buffer casing, 42
—— crank axles, 37–40
—— straight axles, 36, 37
Frame defects, 124, 125
Frames and frame shop, 46–49, 80–82
——, erection of, 80–82, 84–86
——, trammelling, 81

HORNBLOCKS, 48, 87
Hydraulic boiler flanging press, 15–17
—— forging, 41

INSIDE firebox, 20–22
Iron foundry, 27–33

133

134 INDEX

LEAD of the valve, 103–104
Link motion, 78

MACHINING methods, 70
Motion, erection of, 86–87, 91, 92
Moulding for general castings, 29–31

PISTON rods, 73, 74
—— ——, defects and repairs, 131
Pistons and piston rings, 71–73
Plate bending, 13
Port opening of valve, 104–105
Progress of work through the workshops, 4–6

REVERSING lever, effect of position on lead, 105, 106
Running shed repairs, 116, 117

SAFETY valve repairs, 124
Slide-bars, 75, 76
Smoke-box, erection of, 89
Special testing in service, 112
Springs and spring shop, 44, 45, 90
Stamping shop, 43, 44
Stores and costing accounts,

TENDER, 94, 95
Tensile tests, table of, 109
Testing of axles and tyres, 110
—— of materials, chemical tests, 111
—— plant, 115
Tests of materials, bending tests, 109, 110
—— ——, tensile tests, 107–109
Throat plate, 17
Trial trips, 111
Tube defects, 123
—— holes, cracked and oval, 119, 122
Tubes, 23
Tyre fastenings, 63
—— gauges, 64, 66
—— turning and boring mill, 61
Tyres, boring, 62
——, defects and repairs, 127, 128

VALVES and valve gear, defects and repairs, 131
Valve setting, 96–106

WHEEL balancing, 66, 68
—— centres, 60
—— lathes, 64, 65
Wheels and axles, defects and repairs, 126–129
——, placing under engine, 90

Printed by Sir Isaac Pitman & Sons, Ltd., Bath, England
W—(5350A)

A LIST OF BOOKS
PUBLISHED BY
Sir Isaac Pitman & Sons, Ltd.
(*Incorporating WHITTAKER & CO.*)
PARKER STREET, KINGSWAY,
LONDON, W.C.2

The prices given apply only to the British Isles, and are subject to alteration without notice.

A complete Catalogue giving full details of the following books will be sent post free on application.

ALL PRICES ARE NET.

	s.	d.
AERONAUTICAL DESIGN AND CONSTRUCTION, ELEMENTARY PRINCIPLES OF. A. W. Judge	7	6
AEROPLANE STRUCTURAL DESIGN. T. H. Jones and J. D. Frier	21	0
AEROPLANES AND AIRSHIPS. W. E. Dommett	1	9
AIRCRAFT AND AUTOMOBILE MATERIALS—FERROUS. A. W. Judge	25	0
AIRSHIP ATTACKS ON ENGLAND. F. T. Von Buttlar-Brandenfels	1	3
ALTERNATING-CURRENT WORK. W. Perren Maycock	10	6
ALIGNMENT CHARTS. E. S. Andrews	2	0
ARITHMETIC OF ELECTRICAL ENGINEERING. Whittaker's	3	6
ARITHMETIC OF ALTERNATING CURRENTS. E. H. Crapper	4	6
ARMATURE CONSTRUCTION. H. M. Hobart, and A. G. Ellis	25	0
ART AND CRAFT OF CABINET MAKING. D. Denning	7	6
ASTRONOMY, FOR GENERAL READERS. G. F. Chambers	4	0

	s.	d.
ASTRONOMY FOR EVERYBODY. Prof. S. Newcombe	7	6
ATLANTIC FERRY : ITS SHIPS, MEN AND WORKING, THE. A. G. Maginnis	3	0
AUTOMOBILE IGNITION AND VALVE TIMING, STARTING AND LIGHTING, INCLUDING FORD SYSTEM. J. B. Rathbun	8	0
ALTERNATING CURRENT MACHINERY ; PAPERS ON THE DESIGN OF. C. C. Hawkins, S. P. Smith and S. Neville	21	0
BAUDÔT PRINTING TELEGRAPHIC SYSTEM. H. W. Pendry	6	0
CALCULUS FOR ENGINEERING STUDENTS. J. Stoney	3	6
CARPENTRY AND JOINERY : A PRACTICAL HANDBOOK FOR CRAFTSMEN AND STUDENTS. B. F. and H. P. Fletcher	7	6
CENTRAL STATION ELECTRICITY SUPPLY. A. Gay and C. H. Yeaman	12	6
COLOUR IN WOVEN DESIGN : A TREATISE ON TEXTILE COLOURING. R. Beaumont	21	0
ENGLISH-SPANISH COMMERCIAL TERMS. R. D. Monteverde	3	6
COMPRESSED AIR POWER. A. W. and Z. W. Daw	21	0
CONCRETE STEEL BUILDINGS. W. N. Twelvetrees	12	0
CONTINUOUS-CURRENT DYNAMO DESIGN, ELEMENTARY PRINCIPLES OF. H. M. Hobart	10	6
CONTINUOUS CURRENT MOTORS AND CONTROL APPARATUS. W. Perren Maycock	7	6
DESIGN OF AEROPLANES. A. W. Judge	14	0
DESIGN OF ALTERNATING CURRENT MACHINERY. J. R. Barr and R. D. Archibald	30	0
DETAIL DESIGN IN REINFORCED CONCRETE. By E. S. Andrews	6	0
DICTIONARY OF AIRCRAFT. W. E. Dommett	2	0
DETAIL DESIGN OF MARINE SCREW PROPELLERS. D. H. Jackson	6	0
DIRECT CURRENT ELECTRICAL ENGINEERING. J. R. Barr	15	0
DISSECTIONS, ILLUSTRATED. G. G. Brodie	25	0
DIVING MANUAL AND HANDBOOK OF SUBMARINE APPLIANCES. R. H. Davis	7	6
DRAWING AND DESIGNING. C. G. Leland	3	6

	s.	d.
DYNAMO: ITS THEORY, DESIGN AND MANUFACTURE THE. C. C. Hawkins and F. Wallis. In two vols., Each	12	6
ELECTRIC LIGHT FITTING : A TREATISE ON WIRING FOR LIGHTING, HEATING, &c. S. C. Batstone .	6	0
ELECTRO-PLATERS' HANDBOOK. G. E. Bonney .	5	0
ELECTRICAL INSTRUMENT MAKING FOR AMATEURS. S. R. Bottone	6	0
ELECTRO MOTORS. HOW MADE AND HOW USED. S. R. Bottone	4	6
ELECTRIC BELLS AND ALL ABOUT THEM. S. R. Bottone	3	6
ELECTRIC TRACTION. A. T. Dover . . .	25	0
ELECTRIC MOTORS AND CONTROL SYSTEMS. A. T. Dover	18	0
ELECTRIC MOTORS—CONTINUOUS, POLYPHASE AND SINGLE-PHASE MOTORS. H. M. Hobart . .	21	0
ELECTRIC LIGHTING AND POWER DISTRIBUTION. Vol. I. W. Perren Maycock	10	6
ELECTRIC LIGHTING AND POWER DISTRIBUTION. Vol. II. W. Perren Maycock . . .	10	6
ELECTRIC MINING MACHINERY. S. F. Walker .	15	0
ELECTRIC WIRING, FITTINGS, SWITCHES AND LAMPS. W. Perren Maycock	10	0
ELECTRIC WIRING DIAGRAMS. W. Perren Maycock	5	0
ELECTRIC WIRING TABLES. W. Perren Maycock .	5	0
ELECTRIC CIRCUIT THEORY AND CALCULATIONS. W. Perren Maycock	7	6
ELECTRICAL INSTRUMENTS IN THEORY AND PRACTICE. Murdoch and Oschwald . . .	12	6
ELECTRIC LIGHTING IN THE HOME. L. Gaster .		6
ELECTRICAL ENGINEERS' POCKET BOOK. Whittaker's	10	6
ELEMENTS OF ELECTRO-TECHNICS. A. P. Young	7	6
ELEMENTARY GEOLOGY. A. J. Jukes-Browne .	3	0
ELEMENTARY TELEGRAPHY. H. W. Pendry .	4	0
ELEMENTARY AERONAUTICS, OR THE SCIENCE AND PRACTICE OF AERIAL MACHINES. A. P. Thurston.	8	6
ELEMENTARY GRAPHIC STATICS. J. T. Wight .	5	0

	s.	d.
ENGINEER DRAUGHTSMEN'S WORK: HINTS TO BEGINNERS IN DRAWING OFFICES.	2	6
ENGINEERING WORKSHOP EXERCISES. E. Pull	3	6
ENGINEERS' AND ERECTORS' POCKET DICTIONARY: ENGLISH, GERMAN, DUTCH. W. H. Steenbeek	2	6
ENGLISH FOR TECHNICAL STUDENTS. F. F. Potter	2	0
EXPERIMENTAL MATHEMATICS. G. R. Vine		
Book I, with Answers	1	0
" II, with Answers	1	0
EXPLOSIVES INDUSTRY, RISE AND PROGRESS OF THE BRITISH.	18	0
FIELD MANUAL OF SURVEY METHODS AND OPERATIONS. A. Lovat Higgins	21	0
FIELD WORK FOR SCHOOLS. E. H. Harrison and C. A. Hunter	2	0
FILES AND FILING. Fremont, Taylor	21	0
FIRST BOOK OF ELECTRICITY AND MAGNETISM. W. Perren Maycock	6	0
FIVE FIGURE LOGARITHMS. W. E. Dommett	1	6
FLAX CULTURE AND PREPARATION. F. Bradbury	10	6
FUSELAGE DESIGN. A. W. Judge	3	0
GAS, GASOLENE AND OIL ENGINES. J. B. Rathbun	8	0
GAS ENGINE TROUBLES AND INSTALLATIONS. J. B. Rathbun	8	0
GAS AND OIL ENGINE OPERATION. J. Okill	5	0
GAS, OIL, AND PETROL ENGINES: INCLUDING SUCTION GAS PLANT AND HUMPHREY PUMPS. A. Garrard	6	0
GAS SUPPLY IN PRINCIPLES AND PRACTICE: A GUIDE FOR THE GAS FITTER, GAS ENGINEER AND GAS CONSUMER. W. H. Y. Webber	4	0
GERMAN GRAMMAR FOR SCIENCE STUDENTS. W. A. Osborne	3	0
GREAT ASTRONOMERS. Sir R. Ball	7	6
GUIDE TO STUDY OF THE IONIC VALVE. W. D. Owen	2	6
HANDRAILING FOR GEOMETRICAL STAIRCASES. W. A. Scott	2	6
HIGH HEAVENS, IN THE. Sir R. Ball.	10	6
HIGH-SPEED INTERNAL COMBUSTION ENGINES. A. W. Judge	18	0

	s.	d.
HISTORICAL PAPERS ON MODERN EXPLOSIVES. G. W. MacDonald	9	0
HOSIERY MANUFACTURE. W. Davis	9	0
HOUSING PROBLEM. J. J. Clarke	21	0
HOW TO MANAGE THE DYNAMO. A. R. Bottone	2	0
HYDRAULIC MOTORS AND TURBINES. G. R. Bodmer	15	0
INDUCTION COILS. G. E. Bonney	6	0
INSULATION OF ELECTRIC MACHINES. H. W. Turner and H. M. Hobart	21	0
INTRODUCTION TO CHEMICAL ENGINEERING. A. F. Allen	10	6
LEATHER WORK. C. G. Leland	5	0
LEKTRIK LIGHTING CONNECTIONS. W. Perren Maycock		9
LENS WORK FOR AMATEURS. H. Orford	3	6
LIGHTNING CONDUCTORS AND LIGHTNING GUARDS. Sir O. Lodge	15	0
LOGARITHMS FOR BEGINNERS C. N. Pickworth	1	6
MAGNETISM AND ELECTRICITY, AN INTRODUCTORY COURSE OF PRACTICAL. J. R. Ashworth	3	0
MAGNETO AND ELECTRIC IGNITION. W. Hibbert	3	6
MANAGEMENT OF ACCUMULATORS. Sir D. Salomons	7	6
MANUAL INSTRUCTION—WOODWORK. Barter, S.	7	6
” ” DRAWING. ”	4	0
MANUFACTURE OF EXPLOSIVES. 2 Vols. O. Guttmann	50	0
MATHEMATICAL TABLES. W. E. Dommett.	4	6
MECHANICAL TABLES, SHOWING THE DIAMETERS AND CIRCUMFERENCES OF IRON BARS, ETC. J. Foden	2	0
MECHANICAL ENGINEERS' POCKET BOOK. Whittaker's	6	0
MECHANICS' AND DRAUGHTSMEN'S POCKET BOOK. W. E. Dommett	2	6
METAL TURNING. J. G. Horner	4	0
METAL WORK—REPOUSSÉ. C. G. Leland	5	0
METRIC AND BRITISH SYSTEMS OF WEIGHTS AND MEASURES. F. M. Perkin	3	6
METRIC CONVERSION TABLES. W. E. Dommett	2	6

	s.	d.
MINERALOGY : THE CHARACTERS OF MINERALS, THEIR CLASSIFICATION AND DESCRIPTION. F. H. Hatch	6	0
MINING MATHEMATICS (PRELIMINARY). G. W. Stringfellow	1	6
Do. With Answers	2	0
MODERN ILLUMINANTS AND ILLUMINATING ENGINEERING. Dow and Gaster	25	0
MODERN PRACTICE OF COAL MINING. Kerr and Burns. Part 1, 5/-; Parts 2, 3 and 4, each	6	0
MODERN OPTICAL INSTRUMENTS. H. Orford	4	0
MODERN MILLING. E. Pull	9	0
MOTION PICTURE OPERATION, STAGE ELECTRICS AND ILLUSIONS. H. C. Horstmann and V. H. Tousley	7	6
MOTOR TRUCK AND AUTOMOBILE MOTORS AND MECHANISM. T. H. Russell	8	0
MOTOR BOATS, HYDROPLANES AND HYDROAEROPLANES. T. H. Russell	8	0
MOVING LOADS ON RAILWAY UNDERBRIDGES. H. Bamford	5	6
OPTICS OF PHOTOGRAPHY AND PHOTOGRAPHIC LENSES. J. T. Taylor	4	0
PIPES AND TUBES : THEIR CONSTRUCTION AND JOINTING. P. R. Björling	4	0
PLANT WORLD : ITS PAST, PRESENT AND FUTURE, THE. G. Massee	3	0
PLYWOOD AND GLUE, MANUFACTURE AND USE OF, THE. B. C. Boulton	7	6
POLYPHASE CURRENTS. A. Still	7	6
POWER WIRING DIAGRAMS. A. T. Dover	7	6
PRACTICAL EXERCISES IN HEAT, LIGHT AND SOUND. J. R. Ashworth	2	6
PRACTICAL ELECTRIC LIGHT FITTING. F. C. Allsop	6	0
PRACTICAL SHEET AND PLATE METAL WORK. E. A. Atkins	10	0
PRACTICAL IRONFOUNDING. J. G. Horner	10	0
PRACTICAL TESTING OF ELECTRICAL MACHINES. L. Oulton and N. J. Wilson	6	0
PRACTICAL TELEPHONE HANDBOOK AND GUIDE TO THE TELEPHONIC EXCHANGE. J. Poole	15	0

	s.	d.
PRACTICAL ADVICE FOR MARINE ENGINEERS. C. W. Roberts	5	0
PRACTICAL DESIGN OF REINFORCED CONCRETE BEAMS AND COLUMNS. W. N. Twelvetrees .	7	6
PREPARATORY COURSE TO MACHINE DRAWING. P. W. Scott	1	9
PRIMER OF TRIGONOMETRY FOR ENGINEERS. W. G. Dunkley	5	0
PRIMER OF ENGINEERING SCIENCE. E. S. Andrews Part 1, 3s. ; Part 2, 2s. 6d. ; Complete . .	4	6
PRINCIPLES OF FITTING. J. G. Horner . .	7	6
PRINCIPLES OF PATTERN-MAKING ,, . .	4	0
PROPERTIES OF AEROFOILS AND RESISTANCE OF AERODYNAMIC BODIES. A. W. Judge . .	18	0
QUESTIONS AND ANSWERS FOR AUTOMOBILE STUDENTS AND MECHANICS. T. H. Russell . .	8	0
RADIO-TELEGRAPHIST'S GUIDE AND LOG BOOK. W. H. Marchant	5	6
RADIUM AND ALL ABOUT IT. S. R. Bottone . .	1	6
RAILWAY TECHNICAL VOCABULARY. L. Serraillier	7	6
REINFORCED CONCRETE. W. N. Twelvetrees .	21	0
RESEARCHES IN PLANT PHYSIOLOGY. W. R. G. Atkins	9	0
ROSES AND ROSE GROWING. R. G. Kingsley, .	7	6
ROSES, NEW		9
RUSSIAN WEIGHTS AND MEASURES, TABLES OF. Redvers Elder	2	6
SCIENCE OF THE SOIL. C. Warrell . . .	3	6
SIMPLIFIED METHODS OF CALCULATING REINFORCED CONCRETE BEAMS. W. N. Twelvetrees . .		9
SLIDE RULE. A. L. Higgins		6
SLIDE RULE. C. N. Pickworth	3	6
SMALL BOOK ON ELECTRIC MOTORS, A. C.C. AND A.C. W. Perren Maycock	5	0
STARRY REALMS, IN. Sir R. Ball . . .	10	6
STORAGE BATTERY PRACTICE. R. Rankin . .	7	6
STEEL WORKS ANALYSIS. J. O. Arnold and F. Ibbotson	12	6
STRESSES IN HOOKS AND OTHER CURVED BEAMS. E. S. Andrews	6	0

	s.	d.
STRUCTURAL IRON AND STEEL. W. N. Twelvetrees.	7	6
SUBMARINE VESSELS, ETC. W. E. Dommett	5	0
SURVEYING AND SURVEYING INSTRUMENTS. G. A. T. Middleton	6	0
TABLES FOR MEASURING AND MANURING LAND. J. Cullyer	3	0
TABLES OF SAFE LOADS ON STEEL PILLARS. E. S. Andrews	6	0
TEACHER'S HANDBOOK OF MANUAL TRAINING: METAL WORK. J. S. Miller	4	0
TELEGRAPHY: AN EXPOSITION OF THE TELEGRAPH SYSTEM OF THE BRITISH POST OFFICE. T. E. Herbert	18	0
TEXT-BOOK OF AERONAUTICAL ENGINEERING. A. Klemin	15	0
TEXT BOOK OF BOTANY. Part I—THE ANATOMY OF FLOWERING PLANTS. M. Yates.	2	0
TRANSFORMERS FOR SINGLE AND MULTIPHASE CURRENTS. G. Kapp	12	6
TREATISE ON MANURES. A. B. Griffiths	7	6
TRIPLANE AND THE STABLE BIPLANE. J. C. Hunsaker	3	0
TURRET LATHE TOOLS, HOW TO LAY OUT	6	0
UNION TEXTILE FABRICATION. R. Beaumont	21	0
VENTILATION OF ELECTRICAL MACHINERY. W. H. F. Murdoch	3	6
VENTILATION, PUMPING, AND HAULAGE, THE MATHEMATICS OF. F. Birks	5	0
VOLUMETRIC ANALYSIS. J. B. Coppock	3	6
WIRELESS TELEGRAPHY AND HERTZIAN WAVES. S. R. Bottone	3	6
WIRELESS TELEGRAPHY: A PRACTICAL HANDBOOK FOR OPERATORS AND STUDENTS. W. H. Marchant	7	6
WOODCARVING. C. G. Leland	5	0

Catalogue of Scientific and Technical Books post free.

LONDON : SIR ISAAC PITMAN & SONS, LTD.,
PARKER STREET, KINGSWAY, W.C.2

Printed in Great Britain by
Amazon.co.uk, Ltd.,
Marston Gate.